VHDL FOR SIMULATION, SYNTHESIS AND FORMAL PROOFS OF HARDWARE

THE KLUWER INTERNATIONAL SERIES
IN ENGINEERING AND COMPUTER SCIENCE

VHDL FOR SIMULATION, SYNTHESIS AND FORMAL PROOFS OF HARDWARE

edited by

Jean Mermet
Institut Méditerranéen de Technologie

KLUWER ACADEMIC PUBLISHERS
DORDRECHT / BOSTON / LONDON

Library of Congress Cataloging-in-Publication Data

VHDL for simulation, synthesis, and formal proofs of hardware / edited by Jean Mermet.
p. cm. -- (Kluwer international series in engineering and computer science ; SECS 183)
ISBN 0-7923-9253-1
1. VHDL (Computer hardware description language) I. Mermet, Jean P. II. Series.
TK7885.7V48 1992 92-16266
621.39'2--dc20

ISBN 0-7923-9253-1

Published by Kluwer Academic Publishers,
P.O. Box 17, 3300 AA Dordrecht, The Netherlands.

Kluwer Academic Publishers incorporates
the publishing programmes of
D. Reidel, Martinus Nijhoff, Dr W. Junk and MTP Press.

Sold and distributed in the U.S.A. and Canada
by Kluwer Academic Publishers,
101 Philip Drive, Norwell, MA 02061, U.S.A.

In all other countries, sold and distributed
by Kluwer Academic Publishers Group,
P.O. Box 322, 3300 AH Dordrecht, The Netherlands.

Printed on acid-free paper

Printed in the Netherlands

Table of Contents

Preface

The success of VHDL since it has been balloted in 1987 as an IEEE standard may look incomprehensible to the large population of hardware designers, who had never heared of Hardware Description Languages before (for at least 90% of them), as well as to the few hundreds of specialists who had been working on these languages for a long time (25 years for some of them).

Until 1988, only a very small subset of designers, in a few large companies, were used to describe their designs using a proprietary HDL, or sometimes a HDL inherited from a University when some software environment happened to be developped around it, allowing usability by third parties.

A number of benefits were definitely recognized to this practice, such as functional verification of a specification through simulation, first performance evaluation of a tentative design, and sometimes automatic microprogram generation or even automatic high level synthesis.

As there was apparently no market for HDL's, the ECAD vendors did not care about them, start-up companies were seldom able to survive in this area, and large users of proprietary tools were spending more and more people and money just to maintain their internal system.

In the community of Hardware Description Language designers, which I belong to, people knew about several dozens of more or less successful HDL's, some of which appeared in the early sixties: "A formal language describing machine LOgic Timing and Sequencing (LOTIS)" by Schlaeppi, IBM, in 1964; "EPICURE" in an ancestor company of BULL in 1965; "An Algol-like Computer Design Language: CDL" by Chu, University of Maryland, in 1965; "HARGOL, a hardware-oriented Algol language" by Giese at University of Copenhagen; "CASSANDRE language" by myself at University of Grenoble, 1968; "A Digital system Design Language: DDL" by Duley and Dietmeyer at University of Wisconsin, 1969 etc...
Some of these languages had a compiler, a simulator and sometimes a synthesis derivation such as "ALERT: a program to produce logic design for preliminary machine descriptions" by Friedman (IBM, 1966) using APL.

During the seventies, although Japan was rather absent of this game, HDL's proliferated in Europe and in USA. More than 200 happened to be mentioned in a survey! The Conference on Hardware Description Languages and their Applications decided to establish some order in this "tower of Babel", and mandated a group of six specialists, R. Piloty, M. Barbacci, D. Borrione, D. Dietmeyer, F. Hill and P. Skelly to create a "consensus language": CONLAN. Started in 1976, this work was the first international standardization effort in this domain. The papers and reports issued in 1980, 1982, 1983 served as a basis for the development of "third generation" HDL's based on CONLAN, such as CASCADE in 1985 or REGLAN. It was also a source of inspiration for the DoD to issue a call for proposals for VHDL, as well as for the early VHDL designers. CONLAN is also the first attempt to provide a complete semantic definition of a HDL.

Then came VHDL, and suddenly a worldwide dissemination in five years of the concepts going along with HDL's, most of which known for twenty years. There are many explanations. But the first one, in my opinion, is the power of the DoD to finance the first developments at the proper level and to impose its own standards to the U.S. industry (where 90% of the ECAD tools vendors can be found). Immediately after comes the urgent (and rather new) need of designers to exchange design descriptions between users, system manufacturers and vendors, the need also to master the complexity and the size of today 's designs that previous methods could no longer achieve. Then, the community of HDL designers was ready to accept the idea that an existing standard was better than no standard at all, even more as VHDL is also a *good* HDL.

The IEEE process of re-standardization occurring every five years has given to this community the ways to convey to VHDL'92 part of their expectations and some of the good features they had implemented in their own HDL's.

The situation was mature for the emergence of a worldwide standard, which does not mean that, in the future, a few other HDL's will not share this position with VHDL. Because of its dissemination and, may be, because it is not so simple, VHDL will educate designers to the concepts of HDL's which remained the property of a rather small community during 30 years. Thirty years, this is the length of a technology generation, in any domain of technology.

The Euro-VHDL conferences, which I started in Marseille in September 1990, met immediately a significant success, reaching the level of the corresponding American meetings. The authors of this book have been nominated as the outstanding contributors to Euro-VHDL'90 and Euro-VHDL'91 (Stockholm). I am sure that their contributions to simulation, synthesis, formal proofs of hardware, and system level design will constitute an extremely valuable medium for all those who want to deepen their knowledge of VHDL, whether they want to teach it, to use it in design, to anticipate future domains of application of VHDL, or even to develop new design environments around VHDL.

Jean Mermet
Marseille, 15 March 1992

Acknowledgements:

I want to thank warmly Franz Rammig, Alec Stanculescu, Ahmed Jerraya, Alex Zamfirescu who have helped editing this book, Adam Postula who was the excellent organizer of Euro-VHDL'91, the C.E.C. who has been the main contributor to the ECAD standardization activities in Europe, through the ESPRIT project ECIP2, and IMT where I have found the best conditions to work in these matters.

Evolutionary Processes in Language, Software, and System Design

F. Erich Marschner
President, CLSI

For almost a decade now, a vision of better tools for electronic systems design has been becoming a reality. From the minds of a small group of people who first met in 1981, the concept of a standard language for the design and description of electronic systems has blossomed into a language definition – IEEE Standard 1076-1987, the VHSIC Hardware Description Language (VHDL) – and an ever-growing set of tools and methodologies for that language.

The evolution of that vision and concept into a language and applications has involved many people and many ideas. The process by which this evolution has occurred is a classic example of the leveraged growth that is the hallmark of the engineering process. We continually learn from experience, in the process developing a better understanding of the principles underlying a given technique or phenomenon, and eventually applying that increased understanding to the development of new theories, new techniques, and new systems that give us increased capabilities to observe or affect our world. These new capabilities in turn give rise to new experiences, and so the bootstrapping process continues, taking us to ever-increasing heights of understanding.

A practical application of this process is the development of tools that, when completed, allow us to build better tools. A good example of this is the development of compiler technology over the last several decades. In the early days of computer use, programming was done by hand at the machine code level; later, symbolic assembly languages and processors were created that made it possible to generate machine code faster. Still later, high-level programming languages were developed that made possible the generation of assembly language more quickly and more accurately. Today, every computer-literate person understands the role of a compiler in the development of computer programs, but few would consider the compilation process as more than a primitive operation.

J. Mermet (ed.), VHDL for Simulation, Synthesis and Formal Proofs of Hardware, 1–13.

One of the most interesting situations in which this continued cycle of leveraged growth applies is that in which a tool is being developed that in turn can be used to develop future generations of the same tool. For example, consider the standard paradigm of software development, in which source code is compiled to object code, and the object code is in turn linked to create an executable image. But what is the compiler itself? It is the executable image derived from its own source code, so the same paradigm applies to creating the compiler. Now if the compiler is written in the same language that it compiles, the process becomes recursive, and the compiler acquires the ability to compile itself.

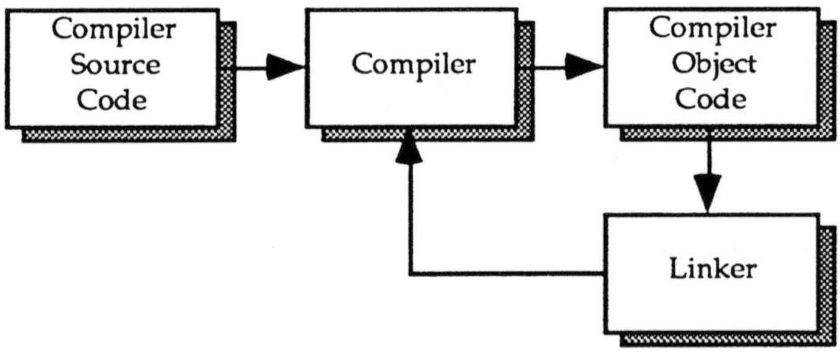

Figure 1. A Self-Compiling Compiler.

On the surface, a self-compiling compiler looks like a high-tech version of the "chicken and the egg" problem — which came first, the executable image or the source code? Of course, the answer is the same for both — just as Nature has spent countless iterations developing chickens that can produce eggs that in turn produce other chickens, software engineers have applied the same bootstrapping process (over a shorter period of time, but with perhaps no fewer iterations) to create an executable program that can produce object code that in turn can be linked to produce the same executable program.

From this point of view, the source code becomes almost secondary; like the DNA that defines the chicken, the source is just a template that defines the executable image. And in fact, just as chickens can evolve over time via genetic change, a self-compiling compiler can evolve over time via changes to its source code.

One of my earliest experiences as a software engineer involved working with a self-compiling Pascal compiler. My job was to help it evolve from the simple, stack-based runtime architecture that it initially employed to a more complex, register-oriented architecture suitable for operating system applications. In order to think clearly about the recursive stages in its evolution, we adopted a notation called the "T-diagram." This notion clearly shows the bootstrapping process, in which each cycle uses an existing executable image of the compiler to produce the executable image for the next cycle.

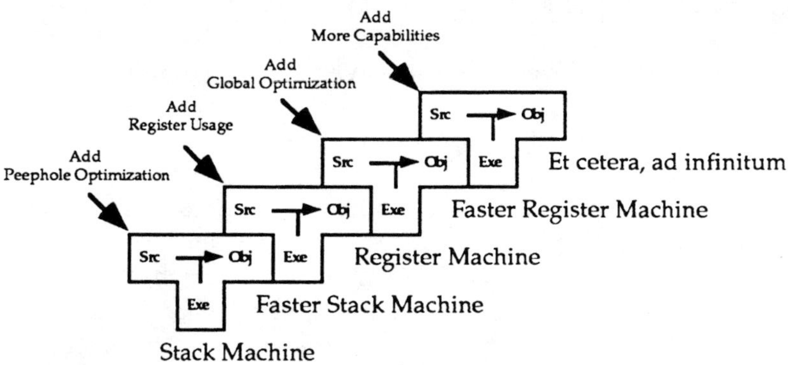

Figure 2. A "T-Diagram" Representation of Compiler Evolution.

The first step involved adding peephole optimization to the code generator. This removed redundant push/pop operations, so the resulting executable was smaller and faster. The next step was much more significant, involving the conversion of stack operations to operations on temporary storage locations, which could then be mapped onto registers in a straightforward fashion. This produced an executable image that employed all of the registers of the underlying host and therefore ran significantly faster. Finally, we applied various global optimization techniques and register allocation optimizations. This resulted in a new executable image that was significantly smaller and faster than the previous one, and nearly an order of magnitude faster than the original, stack-based machine.

This evolutionary process could have continued forever, with each cycle resulting in increased performance in the compiler. However, we had a particular goal in mind, namely, the creation of a compiler to support the development of an operating system written in Pascal. In fact, the operating system required a special dialect of Pascal, developed by Per Brinch Hansen, called Concurrent Pascal. In Concurrent Pascal, one can define concurrent processes that interact via objects, called "monitors", that protect shared resources. A monitor is essentially a critical section in which only one process can execute at a time.

Concurrent Pascal also includes facilities for data abstraction, in the form of the "class" construct. A class is a template for a data structure together with operations on that structure. Classes can be instantiated within processes, monitors, or other classes, just as variables of a given type can be declared. A class, however, encapsulates both data and operations in one definition, thus it supports object-oriented methodologies quite well.

In order to evolve my self-compiling Pascal compiler to the point where it would support the development of an operating system, I first had to teach it to run on the host machine for which the operating system was being developed, and then I had to teach it Concurrent and Sequential Pascal. Once again, I found myself bootstrapping the compiler to increase its capabilities.

This time, I started with an optimized, register-oriented executable, modified its source to generate code for the new target, and then compiled the modified source. This produced a cross-compiler, still running on the original host, but now capable of generating code for a new target. The next step simply involved porting the compiler to the new host by compiling it with the cross-compiler executable and moving the objects to the new host for linking. (The hardest part of this step was figuring our how to transfer binary files on tape from one machine to the other.) Finally, after adding to the rehosted compiler's source code the requisite features to support Concurrent and Sequential Pascal, I used the rehosted compiler to build a Concurrent and Sequential Pascal compiler executable.

That was about as far as I went with the evolution of my self-compiling compiler. The operating system was never built (a casualty of United States Department of Defense budget cuts), although we did implement a file system in Concurrent Pascal that verified the correct operation of the compiler. However, if the operating system had been built, I would have wanted to take the process one more step, by rewriting both the compiler and the linker as Sequential Pascal programs as part of the new operating system. This final transformation would have allowed the operating system to be written in such a way that it could monitor its own source code, and recompile and relink itself whenever its source code changed. Such a system could actually be aware (in a minimal sort of way) that it is processing its own code, which raises a number of interesting philosophical questions. Unfortunately, I was never able to pursue this exercise. Instead, I turned my attention to the design of a language for defining hardware rather than software – VHDL.

The evolution of VHDL over the last decade exhibits the same cyclic nature that characterizes all bootstrapping situations. Throughout, the language has been driven by the need to develop software for automating the electronic design process. Existing applications, and experience with tools that worked well or not so well in those areas, have led to requirements on the language capabilities, which in turn governed the design of the language. Implementation of the language has revealed new issues, and application of the language has led to new requirements.

The concept of a standard hardware description language for electronic design was crystallized at the 2-week long Woods Hole Summer Study on Hardware Description Languages, which occurred in June of 1981. At this workshop, approximately 20 people from industry and academia got together to develop requirements for a hardware description language for use in the U. S. Department of Defense VHSIC program. After considering various existing languages, including Ada, the group concluded that no existing hardware description language or software programming language was sufficient to meet the needs of the VHSIC program, because no existing language met all of the requirements. Instead, the group recommended that a new language be created. During 1981, the Woods Hole requirements were merged with additional requirements from the Ada Steelman document, resulting in a document entitled "Department of Defense Requirements for Hardware Description

Languages." This requirements document became the basis for the initial development of VHDL.

The design of the original version of VHDL took approximately one year, starting in October of 1983. In August 1984, the preliminary design was presented in a design review in Orono, Maine. Reviewers from industry, academia, and the Department of Defense provided many comments to the design team, which resulted in an additional 4 months of design activity. This process culminated in the definition of VHDL version 7.2.

To prove the utility of the language definition, a set of tools were developed to support analysis and simulation of VHDL descriptions. Also, a set of benchmarks were created to help evaluate the utility of the language for documenting hardware. In addition, the Department of Defense required the use of VHDL 7.2 for documenting several chips developed within the VHSIC program.

The focus of this initial round of development was on documentation and simulation. These applications were considered most critical at the time: documentation would support re-procurement of systems after their initial design in such a way that advances in technology could be incorporated into new systems, and simulation would support verification of the design to make sure its functional behavior was correct. Although there was some interest in supporting other aspects of the design process, such as physical design, the Department of Defense Requirements for HDLs made it clear that these were secondary issues. It is interesting to note that there was hardly any mention of synthesis in those requirements, yet synthesis is one of the most prominent applications of VHDL today.

The second cycle in the development of VHDL began in 1986, with the creation of the VHDL Analysis and Standardization Group (VASG) within the IEEE Design Automation Technical Committee's Design Automation Standards Subcommittee. This group was charged with the responsibility for evolving the VHDL 7.2 language definition into a form that would be acceptable to the industry, for commercial application as well as for use in defense contracts. Once again, experience with the application of VHDL 7.2 led to requirements on the language design, and these requirements were addressed over a period of approximately one year. The revised language definition was released in several drafts and was balloted as an IEEE Standard in the summer of 1987. Final approval for the standard language came with the affirmative vote of the standards board on December 10, 1987, thus creating IEEE Standard 1076-1987 VHDL.

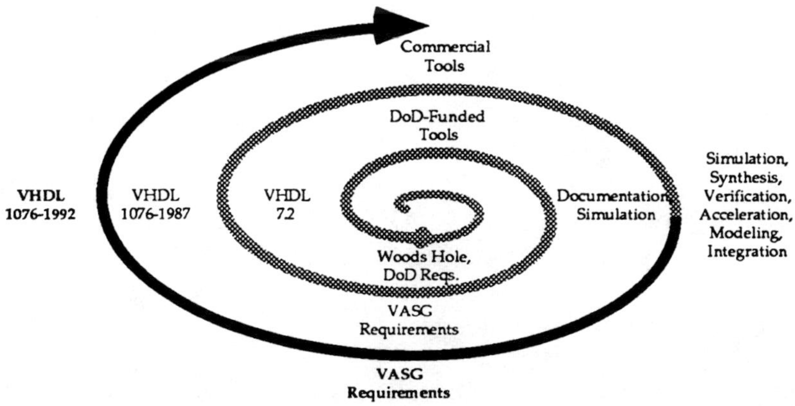

Figure 3. Cycles of VHDL Evolution.

Since then, may developers have set about implementing VHDL 1076. The increased attention to this version of the language, together with the need to integrate it with existing design systems, has led to discussions of subsets and conventions for communicating among tools. For simulation, the question of subsets has been diminishing as more and more tools become complete; subsets for synthesis and for other applications are still being considered.

Simulation is still the primary application for VHDL, and a number of papers in this conference address the subject. Issues such as the development of standard multi-valued logic types and the ability to simulate descriptions at very low or very high levels of abstraction dominate this area. Fault simulation in VHDL has also been receiving some attention. Finally, the need to model analog and mixed analog/digital circuitry in VHDL for mixed mode simulation is becoming important.

Synthesis is the next most significant application area of interest today. In fact, the ability to accelerate the design process through synthesis from VHDL descriptions has apparently become the dominant factor in encouraging the use of VHDL today, and the fact that the U.S. Department of Defense requires VHDL documentation on defense contracts is now less important in promoting the use of the language. This means that the language can now stand on its own as a commercial success, without being subject to the vagaries of defense spending.

There are two general classes of synthesis tools of interest today; logic synthesis tools and behavioral synthesis tools. The former are fairly well understood and have reached the level of commercial implementation; most of the CAE vendors are now offering some form of logic synthesis capability. Behavioral synthesis, or "high-level" synthesis, is still a matter of significant research, most of which is occurring in the universities.

The goal of simulation is to validate a description by checking that it produces the appropriate outputs for a given set of inputs, and to do so for a set of examples that is large enough to indicate a reasonable level of confidence in the design. However there is always the chance that one more test vector will uncover a problem. An alternative method of demonstrating the accuracy of a description is to develop formal proof that the description matches some higher-level specification that is known to be correct. This process of formal verification of a description can in some cases be more cost effective than simulation of the same design. Furthermore, the application of formal methods can often reveal potential trouble spots, even if a formal proof of correctness cannot be obtained. Various researchers are working on the development of formal verification tools for use with VHDL.

VHDL itself is even being used to define a formal framework for the interaction of diverse simulators. A relatively unusual application, this nonetheless demonstrates that the structural concepts and data typing facilities of VHDL can be used in new and creative ways.

These and other applications are the driving forces behind the next cycle of VHDL evolution. As we try to apply VHDL 1076-1987 to these endeavors, we will find that it is capable in some ways and less capable in others. We may find that additional capabilities are required; we may find that slight modifications to existing features would be helpful. The one thing that is certain is that every VHDL user will have an opinion about how the language should change.

As an IEEE Standard, VHDL 1076 must be reballoted every 5 years, and in the process it may be refined to address additional requirements stemming from VHDL applications. Thus the next version of VHDL must be approved by the end of 1992. There is a mechanism being established now to collect the results of applying VHDL 1076 and the implications regarding the need for language change. The VHDL Analysis and Standardization Group is defining special interest groups that will be examining the application of VHDL in various areas, including simulation, synthesis, modeling, formal semantics, and others. These groups will serve to channel opinions about the capabilities of VHDL 1076 into the restandardization process for VHDL. I would encourage all of you who have an interest in affecting the evolution of VHDL to become actively involved in the VASG and in the special interest groups that will be evaluating the language during this next cycle.

After VHDL 1076-1992 is defined, what will be in store for VHDL and VHDL applications? It is impossible to know for certain, but various trends are evident, and perhaps we can hazard a guess about which application areas will see the most activity.

Synthesis will clearly dominate the field in the next few years. The productivity gains to be obtained from the application of synthesis techniques are substantial. Logic synthesis tools are already becoming widespread, and behavioral synthesis will soon be leaving the university research labs to see the light of day in industrial applications.

Verification also seems to be a promising application area, given the advantages it offers over simulation. Although a formal semantic model for VHDL may be required before verification tools become widespread, the ability to be certain that a design is correct far outweighs the ability to believe, with a 95% confidence level, that it is correct — especially when the design is part of a missile system or critical health care system. And, if that can be achieved with less expenditure of computer time and human effort, the verification will quickly replace simulation as the preferred means of validating a design.

Another area in which I expect VHDL applications to develop is that of test. Test pattern generation from VHDL is already being considered. But what I am looking forward to is the development of analytical and diagnostic tools that can monitor the operation of hardware and compare it to the VHDL specification in order to detect errors that develop over time. I expect that verification techniques will be adapted to this end; instead of comparing a VHDL description with its specification, such a tool would compare the actual hardware operation with the specification, using structural information to localize a fault.

Integration of analysis, diagnosis, synthesis, and verification tools into a coherent system will also provide new opportunities. Synthesis and verification are an obvious combination, since it is important to make sure that the results of synthesis match the input specification. Similarly, analysis tools may be combined with synthesis to provide feedback on the results of the synthesis process, perhaps with the effect of changing various control parameters and re-synthesizing in order to achieve a more efficient result. Such combinations for cooperating applications will bring more power to electronic design automation, just as the combination of simulation and synthesis is today.

Finally, high-level synthesis capabilities will promote the implementation of many such design tools directly in hardware for increased performance. Just as we have hardware simulation engines today, we will see hardware verification engines in the future. This implies a need for high-speed, parallel access to design data, so perhaps a VHDL library system and corresponding analysis tools will end up in hardware. Other applications may also benefit from hardware acceleration.

So where does all of this lead? Again, one shouldn't place too much emphasis on the results of speculation, but I am inclined to believe that the systems we develop a decade from now will be as unbelievable today as 100 MIP workstations on a desktop were unbelievable 10 years ago. Thus we may well see some of the following in the coming years:

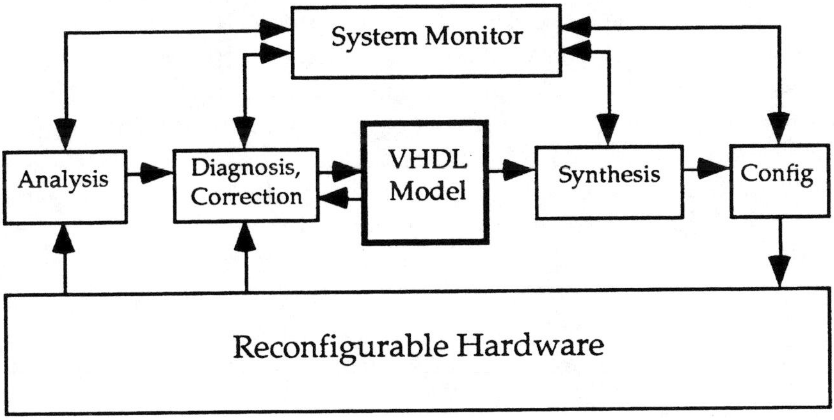

Figure 4. A Self-Modifying System.

- Self-Diagnosing Systems that incorporate a VHDL model of the hardware together with a subsystem to monitor the hardware for errors in real-time; a diagnostic subsystem would locate faulty components using VHDL structural information as a guide. This is hardly surprising; the approach is just a natural extension of Built-In Self-Test techniques.

- Self-Modifying Systems, in which the previously mentioned diagnostic subsystems are augmented with the ability to correct errors via built-in synthesis facilities. This might be realized by generating not layout data, as in a conventional synthesis tool, but rather configuration data for Field-Programmable Gate Arrays, a technology that is already available. In fact, such a system could potentially even "diagnosis" bottlenecks in the design and make design changes prior to re-synthesis. Consider how useful this kind of self-maintenance capability would be for remote systems such as the Magellan spacecraft, which started malfunctioning a few weeks ago as it entered orbit around Venus - 40 million kilometers or more from the nearest maintenance crew.

It should be noted that such analysis, diagnosis, and synthesis capabilities may well be embedded in the VHDL model itself and therefore recursively implemented in the configurable hardware. Here we have just the kind of system I wanted to build in Concurrent Pascal many years ago, but now centered around "compilation" of VHDL to hardware. But this raises a question, one that is actually a corollary of the chicken-and-egg question: which really represents the design – the VHDL description or the hardware realization? In some sense, the VHDL description has now become a template for the entire system: the DNA of EDA. Although the hardware realization will almost certainly process data more quickly, the information content of the VHDL and the hardware is the same.

In 1982, Tracy Kidder published a book entitled, "Soul of the New Machine," in which he recounted the efforts of a team of computer engineers at Data General who developed the Eclipse minicomputer. Their efforts involved countless hours sketching circuits, breadboarding, and working with logic analyzers watching microscopic time intervals pass. At one point, one of the engineers quit his job on the spur of the moment, leaving a note saying that he was going to Vermont to live in a commune and that he would no longer consider any unit of time shorter than a season. (For those of you who are interested, that is about 2^{83} femtoseconds, in VHDL terms.)

The point of the book was that a great deal of human effort went into the development of such a machine, hence the "soul" of the machine was derived from those of its builders. However, if we ever reach the point at which a machine can be built that modifies itself, in essence re-creating itself each time a modification is applied, then in some sense the VHDL model at the heart of the machine will have become its soul. The model will provide the machine with an internal self-image, a prerequisite for becoming aware of itself.

Are self-aware machines possible? I don't know. Perhaps we will need a new kind of Turing Test for this — can a self-modifying machine determine whether it is regenerating itself or generating hardware for some "other" design? Perhaps we will one day build machines that can pass such a test — and if so, the Association for Computing Machinery may have to become the Association *of* Computing Machinery. Only time will tell.

It is interesting to note that the process/monitor structure of Concurrent Pascal is very similar to the structure of VHDL, in which processes communicate via signals. In a sense, a signal is like a special kind of monitor that is tuned for efficient modeling of time-varying sequences of values passing between processes. The analogy can be seen especially when a signal requires a resolution function to resolve multiple inputs and generate a single output. Similarly, the notion of a "class" in Concurrent Pascal parallels that of a "package" in VHDL, with the added benefit of instantiability. It is interesting to conjecture how VHDL would have turned out had the design of the language been based upon Concurrent Pascal rather than Ada.

The runtime environment of Concurrent Pascal is also similar to that of a VHDL simulator. An underlying kernel provides the necessary process scheduling services, and the processes (and monitors) making up the system are statically elaborated, just as in VHDL. One difference is that Concurrent Pascal allows other programs, written in Sequential Pascal (which is essentially the same as standard Pascal with minimal file I/O capabilities), to be loaded into buffers and executed. This capability is necessary, of course, in an operating system.

Part I

SIMULATION

Timing Constraint Checks in VHDL - a comparative study[1]

Fuhong Liu[2]
and Adam Pawlak

Gesellschaft für Mathematik und Datenverarbeitung (GMD)
D-5205 St. Augustin, Germany

1. Introduction

A broad scope and flexibility of the standard hardware description language VHDL enhance on one hand the versatility of its applications. On the other hand, it results in severe problems in its practical use and hinder the acceptance in industry. A challenge faced by most designers is a lack of methodology to effectively apply the broad capabilities of VHDL in a design process. A typical question is, how to model timing constraint checks in VHDL to satisfy certain requirements of consistency, compactness and portability while showing good performance in memory usage and processing time.

In this paper, we address the above mentioned problem of timing constraint checks. Timing checks are an important part of a definition of a standard component model (SCM) [4]. A procedure call is proposed as convention for timing constraint checks due to its advantageous features over other styles. Section 2 provides an overview of a standard component model and comments on how a timing constraint check behaves as an indispensable part of the whole coding practice in a SCM. In Section 3, a classification of constraints is given, followed by a discussion of the capability of VHDL

[1]The paper has been presented at EURO-VHDL, Stockholm, Sept. 1991.
[2]Currently with Beijing IC Design Center, Beijing, China.

J. Mermet (ed.), VHDL for Simulation, Synthesis and Formal Proofs of Hardware, 17–32.

in timing constraint checks, then solutions to a case example – waveform checks by employing different structures – are demonstrated. The comparison of different structures with certain features is subsequently provided. Finally, we shortly discuss the application of VAL to the specification of timing constraints.

2. Standard Component Model (SCM)

Imposing some constraints on the application of VHDL may help users in their design process and speed-up the proliferation of design experience. Many groups have, in fact advanced along this direction and worked on the definition of a standard component model [1; 4; 9]. These include:

- proposals towards a consensus on a multi-valued logic system and associated operations to make models interoperable,
- consistent and effective checks of timing constraints and handling of errors to ensure design reliability,
- provision of an external technology related information to make models portable,
- development of timing control functions to make a model user-modifiable and -selectable,
- formulation of naming conventions and commenting models in a readable way.

A SCM in VHDL may constitute several code blocks (Figure 1). An initialization part is allocated to calculate those parameters which are independent of the simulation time in order to minimize the execution time. The technology related timing information includes backannotation delays (input line delays and load delays), propagation delays, etc. A flexible selection of various cases (maximum, minimum and typical) is provided to accurately model the timing behaviour. Some application conditions, for example, a voltage and a temperature which affect the delay parameters, are also included. To make a model more controllable, some control parameter flags and modes are identified in a generic part. A profound discussion of various elements of a SCM definition may be found in [3]. In the following sections, we concentrate on the timing constraint checks only.

3. Timing Constraint Checks

3.1. Classification of Timing Constraints

To ensure a system works properly, a relationship between signals must satisfy certain conditions. These conditions are reflected as constraints which can be classified as follows:

```
entity dfb111f is

 generic(--technology related information
        --application conditions
        --flags
        --modes);

 port(----);

end dfb111f;

architecture beh of dfb111f is

 --declarative part

begin

 --initialization part

 --timing part (delay handling,
 --error processing, constraint checks)

 --function description part

end beh;
```

Fig. 1. A general structure of a SCM.

- Timing constraints
  They are defined as a set of relationships in time domain on input and internal signals of a component. Those include:
  - setup (data signal(s) must be stable for a certain duration of time before clock signal makes its transition),
  - hold (data signal(s) should keep its value(s) for some time after clock makes its transition),
  - waveform (this includes high and low pulse width, falling and rising transitions), and
  - spikes.
- Requirements on data validity
  This kind of constraints represents a check for the correctness of applied data. For example:
  - invalid data (set='1' and reset='1' simultaneously at D Flip-Flop inputs),
  - memory address out of range.

3.2. Techniques for Handling of Timing Constraints

Due to the flexibility of VHDL, timing constraints can be handled in various ways. In this part, we discuss the capability of VHDL for checking constraint violation and present some examples of checks realized using different VHDL mechanisms.

3.2.1. Assertion Statements for Timing Constraint Checks

In VHDL, timing constraint checks are implemented by assertion statements. The general form [8] of an assertion statement is:

assertion boolean expression
report error message
severity severity level;

When an assertion statement is executed, the boolean condition is evaluated. If the condition (timing constraint) is violated, namely, the boolean condition has the FALSE value, then a message will be reported and the simulation process will be interrupted according to the predefined severity level (note, warning, error, failure).

To evaluate the relationship of signals, it is necessary to identify the transition time (slot) of a signal under concern. The attribute of a signal in VHDL provides the simplest method to test for the signal's transitions. Signal attributes are classified into two types [10]:

- signal related attributes (e.g. s'stable, s'delayed, s'transition),
- signal valued attributes (e.g. s'active, s'event, s'last_value, s'last_event, s'last_active).

It is required that the prefix of a signal attribute should be a static signal. So s'stable'event is allowed but s'event'stable is not allowed because s'event returns a function value.

One signal transition can be represented in a number of ways corresponding to different configurations of above attributes. Following are the examples of checks for the clock transition, from low to high, defined using signal attribute:

```
- clock'event and clock='1'
- not clock'stable and clock='1'
- clock'last_value='0' and not clock'stable
```

Note that a signal valued attribute is disallowed within a subprogram [8]. That means STABLE, QUIET, DELAYED are excluded from a procedure body.

```
entity dfb11lf is
 generic(lib_name,dev_name:string);
 port(d,ck,cln: in t_state;
      q, qn : out t_state:='x');
begin
 clk_check:
 waveform_check(clk,
               T_min_hl,T_max_hl,
               T_min_lh,T_max_lh,
               pw_hi_max, pw_hi_min,
               pw_lo_max,pw_lo_min,
               warning);

end dfb11lf;
```

Fig. 2. A waveform check with a concurrent procedure statement.

3.2.2. Various Styles for Timing Constraint Checks

Timing constraint checks may be expressed in VHDL using different statements in various configurations. Additionally, they may be placed either in an entity declaration or in an architecture body part. But all varieties are based on three primary VHDL statements:

- concurrent assertion statement,
- procedure call statement, and
- process statement.

We have experimented with these structures verifying sophisticated waveforms. Complete waveform checks include tests of low and high pulse widths, as well as rising and falling transition periods. For each check, we considered their maximum and minimum values. In the following, we summarize the advantages (+), as well as disadvantages (−) of application – specific VHDL structures for timing checks.

Procedure call statement: A procedure call statement used for timing checks has a general form:

Timing_check (signals, constraints, messages, severity level);

Figure 2 is an example to check the waveform of a clock signal in a D Flip-Flop. The corresponding procedure body is in the appendix. The procedure call applied for timing checks demonstrates that:

++ Different entities can share the same procedure because a procedure call is embedded in a package body. By storing a package in a common library, an easy access is guaranteed.

+ Complex codes of timing checks are hidden from designers. This makes a component model more readable and also reduces an error prone work.
o Miscellaneous items checking can be implemented in one procedure call statement.
o A procedure can be applied in different situations by using a default value. If a constraint gets a default value, the process will turn to next step without checking this constraint. This kind of procedure is more flexible to satisfy various applications.
o A flexible control of a simulation process can be reached by elaborating a formal parameter with a user-specific severity level.
– There is a limit on a number of attributes (signal valued attribute) used in a procedure body.

<u>**Concurrent assertion statement:**</u> A concurrent assertion statement has a simple form as showed above. The problem solved by a procedure call statement is handled here by a set of concurrent assertion statements (Figure 3). Experiments using concurrent assertion statements for timing constraint checks lead to the following conclusions:

++ Its simple and flexible form can be taken in some special constraint checks, for example, in checking the mismatching of signals.
– Verbose in practical use. For each component, we need to write a set of codes for checking constraint violation. In most cases, this is not necessary because many components need to check the same kind of timing constraint.
– Can't deal with complex constraint checks. An example is that for checks of timing constraints on a composite signal, sometimes we want to know not only whether the composite signal violates a constraint but also which special subelement plays a role in this violation.
– Can't be shared by different users and can't ensure consistency when designing a large system.

<u>**Process statement:**</u> An alternative solution to the above checks is shown in Figure 4 by employing a process statement. Our experiments show that:

++ Since a process statement can be composed of simple assertion statements, procedure call statements, and other VHDL statements, it is competitive for complex constraint checks. Principally, all timing constraint checks of one entity can be implemented in one process statement. However, this will lead to low simulation efficiency because the change of each signal on the sensitivity list will activate the process, resulting in evaluations of all statements within the process. In most cases, this is unnecessary.
– It exhibits the same drawback as a concurrent assertion statement, namely, a source code can not be shared by different entities.

```
entity dfb1l1f is
  ....
begin

---rising edge transition check
assert not (clk'delayed(t_min_lh*ns)'event
            and
   clk'delayed(t_min_lh*ns)'last_value='0'
            and (clk'last_event<t_min_lh*ns))
 report "rising transition is too short"
 severity warning;


assert not (clk'delayed(t_max_lh*ns)'event
             and
    clk'delayed(t_max_lh*ns)'last_value='0'
             and (clk='x'))
 report "rising transition is too long"
 severity warning;

 ---pw_hi pulse width check
     ...
 ---falling_edge transition check
      ...
 ---pw_lo pulse width check
      ...

end dfb1l1f;
```

Fig. 3. A waveform check with a concurrent assertion statement.

3.3. Comparison of Structures and Features

3.3.1. Scope of Application

A concurrent assertion statement is the only one, which can not treat complex constraint violation checks. For example, the subelement of a composite signal violating constraint which in some cases needs to be announced, can not be picked out by this means. Generally speaking, complex constraint checks can be handled by a proper programming within a procedure or a process. An application of a procedure call for checking setup constraint of a composite signal is shown in Figure 5.

3.3.2. Consistency

Consistency is a prerequisite to ensure the correctness of simulation results. In respect to that, a procedure call fits best if it is properly used. It reduces

```
entity dfb1l1f is
 ....
 clock waveform check:
 process(clk)
   ...
 begin

  wait until clk='1' and clk'event;
  t1:=now;
 ---rising_edge transition time check
   assert (now - rising_at>=t_min_lh*ns) or
          (t_min_lh*ns=0ns)
    report "rising transition is too short"
     severity warning;
    ...
 ----pw_hi pulse width check
    ...
 ----falling transition time check
    ...
 ----pw_lo pulse width check
    ...
 end process;

end dfb1l1;
```

Fig. 4. A waveform check with a process statement.

the possibility of making an error in handling complex constraint checks and its consistent form makes the model more readable.

3.3.3. Portability

A procedure call statement proves its superiority among the three structures discussed here. This is due to advantages of a package in which a procedure body is embedded. By placing a package in a common library, each user can access it and each entity can share it. In contrast, a concurrent assertion statement and a process statement are assigned to one current entity.

3.3.4. Compactness

Our comparison takes into account all phases of a component model development starting from a model's source code to its simulation run. This includes:

– VHDL source code,

```
procedure  setup_check(signal data:t_vector;
    signal clk:t_state; setup_value:time;
    severity_sel:severity_level:=warning) is
 variable subelement:string(1 to 4):="0000";
 variable last_change_at:time:=0ns;
 variable data_last_val:t_vector(data'range);
begin
a:loop   wait on data, clk;
 if data'event then
  for i in data'range loop
   case data(i) is
    when 'x' =>case data_last_val(i) is
       when 'x' => null;
       when others =>last_change_at:=now;
       subelement:=f_string(i);
       data_last_val(i):='x';  end case;
  -- case when '0' ...; when '1' ...;

 if clk'event and clk='1' then
  assert now- last_change_at>=setup_value
   report "failure on subelement"&subelement
 end if;  end loop;
end setup_check;
```

Fig. 5. A setup check for an array signal.

- IVAN (Intermediate VHDL Attributed Notation) which is produced by the Analyzer,
- Object file and C definition file; produced during model generation,
- Kernel – an executable simulation model produced by Build, which links separately model-generated units with the core, and
- Simulation Run containing the signal histories after execution of a simulation model.

During our experiment, we found that description styles may take less effect on a source code size and on an IVAN size; however, they play an significant impact on an object file, a kernel and a run size of a model. In fact, using signal valued attribute, we have to assign more memory space. For example, if we describe the transition of a signal clk by *not clk'stable and clk='1'* instead of *clk'event and clk='1'* when we check a setup constraint, the object file size increases much faster with respect to the number of concurrent assertion statements. Figure 6 gives the relations of signal valued attributes, number of assertion statements and memory sizes of object files.

Tables 1 to 3 provide performance data for a setup check, a hold check and a waveform check. Those values are measured based on 20 loops of iden-

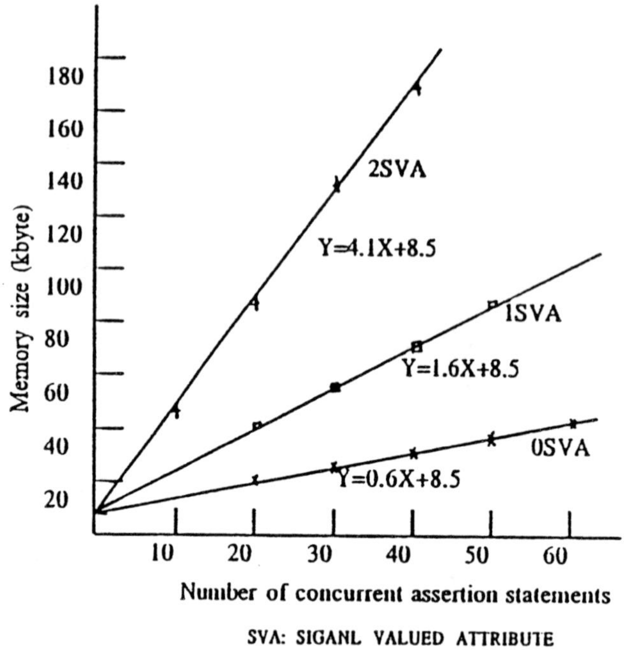

Fig. 6. An influence of a signal valued attribute on a memory size.

tical checks for setup and hold checks to demonstrate the differences clearly. We employed optimized descriptions to give an adequate comparison which means we used as little as possible signal valued attributes in our examples. Unfortunately, for the concurrent assertion statement, it was impossible to represent the transition without using signal valued attribute in hold check and waveform check. It is clear that a procedure call takes the least memory in a VHDL source code and an IVAN code among the three structures. Only for the setup check, it occupies more space in an object file. As more constraint checks are involved in a large system design, the procedure call will show its advantages, leading to a simple and compact source code.

Table 1. Memory Size for HOLD Check (unit:kbyte)

	concurrent assertion	*process*	*procedure call*
VHDL code	4.5	5	2
IVAN	26.5	26.5	17.5
Object File	49	29.5	29.5
Kernel	54.5	41	39.5
Simulation Run	473.5	229	254.5

Table 2. Memory Size for SETUP Check (unit:kbyte)

	concurrent assertion	*process*	*procedure call*
VHDL code	3.5	4	2
IVAN	22	22	17.5
Object File	22	21.5	29.5
Kernel	35	35	39.5
Simulation Run	197.5	197.5	229

Table 3. Memory Size for WAVEFORM Check (unit:kbyte)

	concurrent assertion	*process*	*procedure call*
VHDL code	2.5	2.5	1
IVAN	17.5	17.5	14.5
Object File	41	14.5	11.5
Kernel	36.5	29.5	16
Simulation Run	209.5	98.5	110.5

Table 4. Simulation Time Comparison (unit:second)

	concurrent assertion	*process*	*procedure call*
HOLD Check	33.11	27.06	29.07
SETUP Check	26.59	26.43	29.02
WAVEFORM Check	14.10	12.27	12.86

3.3.5. Simulation Efficiency

Using concurrent assertion statement for timing constraint checks will introduce additional attribute signals in a process. In fact, 16 such signals are added in our WAVEFORM check example. The more signal valued attribute signals appear in the description, the more times the assignment statement is activated. This will cause extra evaluation of a boolean condition. We can not check the events because our Intermetrics simulator is not equipped with this facility. From the simulation results, it seems that more events are involved due to the introduction of signal valued attributes. In addition, when we check the transition of a clock, we do not want those statements for checking pulse width evaluated. Unfortunately, it is impossible that these two concurrent assertion statements are executed in parallel because they are all sensitive to the clock. Table 4 summarizes the simulation results for three kinds of constraint checks. The simulation time has been measured in the cpu seconds on a VAX/3200. An execution of a procedure call takes less time than that of a concurrent assertion statement and more than that of a process statement.

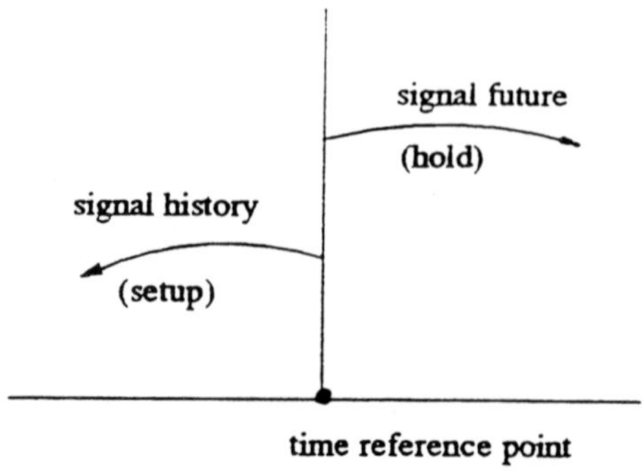

Fig. 7. A classification of timing constraints according to their time precedences.

4. Application of VAL for timing specification

In this part, we discuss from a language point of view some requirements for the extension of VHDL in the timing specification domain. The advantages of VAL will be shortly analyzed.

Although we have classified timing constraints with respect to their physical features, the constraints can also be viewed according to their relative time precedences, namely, from a time reference point tracing a signal's history (typically, setup) or predicating a signal's future (typically, hold time) (Figure 7). VHDL supports directly the first type of constraint checks by employing signal related attributes (S'last_event, etc.). A signal related attribute S'last_event returns the time period. This is the time elapsed since the last event on the signal. Unfortunately, there are no well defined constructs for postulating the stability of a signal in the future time interval. One possible approach is to introduce an additional signal – a signal valued attribute (for example, s'delayed(t)). However, this technique is not efficient both in coding and simulation, as we have argued above. Furthermore, it is not easy to comprehend, since designer's intentions are expressed not explicitly.

The inception of VAL [11; 12] fills this gap by introducing a qualified expression. The syntactic form of this expression in general, can be represented as: "**e1** during **e2**", where **e1** is a boolean expression and **e2** is an integer or a real or a time interval. The qualified expression is suitable to model the stability of a signal in a time interval. Another important feature is a guarded assertion statement. It provides a flexible way for building assertion constructs, and in addition is very useful for the synchronization of assertion statements to achieve high simulation speed.

As an example, we employ VAL to describe the timing dependencies of a complex waveform. Figure 8 shows a write cycle of RAM 2149. The timing constraint checks in VAL are presented in Figure 9. Note that we

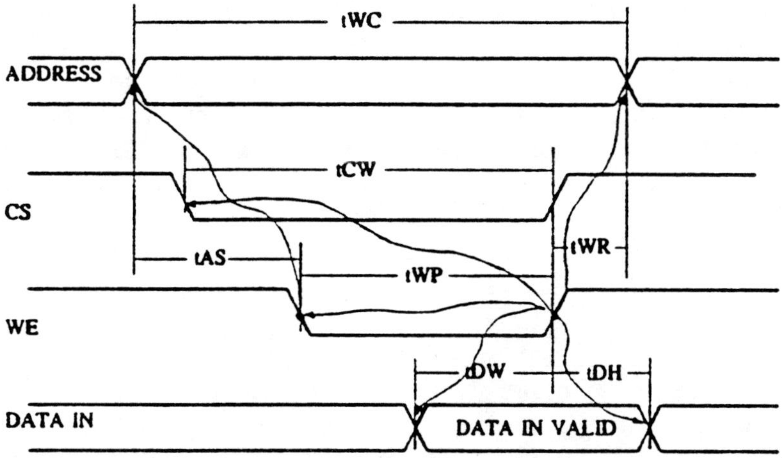

Fig. 8. A write cycle of RAM 2149.

```
entity ram is
 generic(tWC, tCW, tAW, tAS, tWP, tWR,
         tDW, tDH, tWZ, tOW : time := 0 ns);

 port(a : in t_vector; -- address
      di : in t_vector; -- data in
      ncs, nwe : in t_state;
      do : out t_vector); --data out
 begin

--|when nwe'changed('0') then
--| finally (not a'event) during [-tAS,0ns];
--|end when;

--|when nwe'changed('1') then
--| finally (not a'event) during [-tAW,0ns];
--| finally (not ncs'event) during [-tCW,0ns];
--| finally (not nwe'event) during [-tWP,0ns];
--| finally (not di'event) during [-tDW,0ns];
--| finally (not a'event) during [0ns,tWR];
--| finally (not di'event) during [0ns,tDH];
--|end when ;

end ram ;
```

Fig. 9. Timing dependency checks with VAL.

have selected two time reference points, i.e., control signal *nwe* at its rising edge and falling edges respectively. All checks are based on them. With the guarded structure "when ... end when", we group constraint checks into two processes. "finally ..." in code represents a loose assertion, which asserts only the last simulation cycle of the time point in which the assertion process becomes active. This type of assertion reflects most designers' intention, since in most cases, they are more interested in the final stable value of a signal.

The current VAL version does not support annotations in a package. We feel that sometimes it is more convenient to place some timing dependencies checks in a procedure for later use and for maintenance. Another possible extension may be to introduce an additional signal related attribute *S'pw(st)* for the specification of a signal pulse width, where *S* represents a signal, *pw* pulse width and *st* a state of a signal. Since *S'pw(st)* only relates to one signal's history, it can be easily implemented. This extension promotes, clarity on the specification of timing constraints.

5. Conclusions

A procedure call is proposed as a convention for timing constraint checks. Our experiments have proved that this structure satisfies the requirements of consistency, portability, compactness and shows also a favorable tradeoff between a memory size and an execution time.

The paper also shortly discussed the application of VAL for timing constraint checks. The qualified expression of VAL shows its adequacy in the representation of a signal's postulated stability. VAL being the extension of VHDL has some competitive language features in timing specification domain. They should be taken into account in the language design phase of the VHDL restandardization.

6. Acknowledgements

Let us take a chance to express our gratitude to members of the VAL development group, i.e. Alec Stanculescu and Larry M. Augustin for providing us with the VAL code and the support tool. Many thanks also to Intermetrics for granting us with the university licence for their VHDL platform.

References

[1] Coelho D.V.: *The VHDL Handbook*, Kluwer Academic Publishers 1989.

[2] Armstrong J.R.: *Chip Level Modeling with VHDL*, Prentice-Hall, 1989.

[3] Dubois J-L, Liu F., Pawlak A.: *Standard Component Modeling with VHDL*, Proc. of ASIM'91. Hagen, Germany, Sept. 1991.

[4] EIA: *EIA Commercial Component Model Specification SP-229.*

[5] Joseph S. Lis.: *VHDL Design Representation in the VHDL Synthesis System (VSS)*, Technical report, University of California, 1989.

[6] Aylor J.H., Waxman R., Scarratt C.: *VHDL—Feature Description and Analysis*

[7] EIA: *Data Item Description Entitled Electronic Hardware Description of the VHSIC Hardware Description Language (VHDL) Documentation*, Identification Number DI-EGDS-80811.

[8] IEEE: *IEEE Standard VHDL Language Reference Manual*, IEEE, Inc., NY, March 1988.

[9] IEEE: *IEEE Design & Test of Computer*, 1990.

[10] Lipsett R, et al.: *VHDL: Hardware Description and Design*, Kluwer Academic Publishers 89.

[11] Luckham D. C., et al: *The Semantics of Timing Constructs in Hardware Description Languages*, Proc. of ICCD '86, page 10 – 14.

[12] Luckham D. C., et al: *Hardware Design and Simulation in VAL/VHDL*, Technical report of Stanford University, May 1990.

Appendix

```
procedure waveform_check(signal clk:t_state;
   constant T_min_hl,T_max_hl:time:=0 ns;
   constant T_min_lh,T_max_lh:time:=0 ns;
   constant pw_hi_max,pw_hi_min:time:=0 ns;
   constant pw_lo_max,pw_lo_min:time:=0 ns;
   constant severity_sel: severity_level
                              :=warning) is
 variable rising_at,t1:time:=0 ns;
 variable falling_at,t0:time:=0 ns;
begin
a:loop wait until clk='1' and clk'event;
----rising_edge transition time check
 if now>0 ns then
  assert (now - rising_at>=t_min_lh)
    or (t_min_lh=0 ns)
   report "rising transition is too short"
  severity severity_sel;
  assert (now - rising_at<=t_max_lh)
    or (t_max_lh=0 ns)
   report "rising transition is too long"
```

```
    severity severity_sel;
    end if;    t1:=now;
    wait until clk'event and clk'last_value='1';
    falling_at:=now;
  ----pw_hi pulse width check
    assert (now - t1>=pw_hi_min)
       or (pw_hi_min=0 ns)
     report "pw_hi is too short"
       severity severity_sel;
    assert (now - t1<=pw_hi_max) or
       (pw_hi_max=0 ns)
     report "pw_hi is too long"
      severity severity_sel;
     wait until clk'event and clk='0';  t0:=now;
    ----falling transition time check
    assert (now - falling_at>=t_min_hl)
       or (t_min_hl=0 ns)
     report "falling transition is too short"
      severity severity_sel;
    assert (now - falling_at<=t_max_hl)
       or (t_max_hl=0 ns)
     report "falling transition is too long"
      severity severity_sel;
     wait until clk'event and clk'last_value='0';
      rising_at:=now;
    ----pw_lo pulse width check
     assert (now - t0>=pw_lo_min)
      or (pw_lo_min=0 ns)
     report "pw_lo is too short"
      severity severity_sel;
     assert (now - t0<=pw_lo_max)
      or (pw_lo_max=0 ns)
      report "pw_lo is too long"
       severity severity_sel;
   end loop;
end waveform_check;
```

Using Formalized Timing Diagrams in VHDL Simulation

M. DUFRESNE, K. KHORDOC, E. CERNY

Dép. d'IRO, Université de Montréal
C.P. 6128, Succ. A, Montréal, Qué., H3C 3J7 Canada

1. Introduction

There is presently a need for timing verification tools that address system level design. The designers spend a large amount of time writing system inputs and checking functions to verify timing and functional specifications. We propose a new tool that facilitates the timing verification of complex systems in the context of behavioral simulation. Because timing behavior is often described using timing diagrams and because this kind of diagram is well understood by hardware designers, our tool uses hierarchical formalized timing diagrams [1] to capture specifications. The tool extracts timing specifications from these diagrams to automatically perform stimulus generation and response validation during the simulation run-time. Our tool uses a new algorithm that traverses the constraint graph hierarchy during simulation. In the rest of this paper we call our system DSGRV, for Dynamic Stimulus Generation and Response Validation.

The paper is structured as follows: Section 2 presents an overview of our system. Section 3 briefly introduces the model. Section 4 and 5 present the VHDL implementation including the DSGRV process. Section 6 contains experimental results and, finally, Section 7 concludes the presentation.

2. System organization

Our system is divided in two main parts as shown in Fig. 1. The first part of the system is written in LISP and contains a waveform editor and a VHDL code generator. The waveform editor (*Shadow*) is from Bell-Northern

J. Mermet (ed.), VHDL for Simulation, Synthesis and Formal Proofs of Hardware, 33–42.

Research Ltd. and we developed the VHDL code generator as an extension of *Shadow*. We use the editor to specify hierarchical timing diagrams and the VHDL generator to automatically generate two VHDL modules. The first generated module is the DSGRV entity and architecture and the second corresponds to the top level test bench entity and architecture that establishes the communication between DSGRV and the circuit. The second part of our system executes in a VHDL environment and contains our VHDL packages, the synthesized modules and description of the circuit to be simulated.

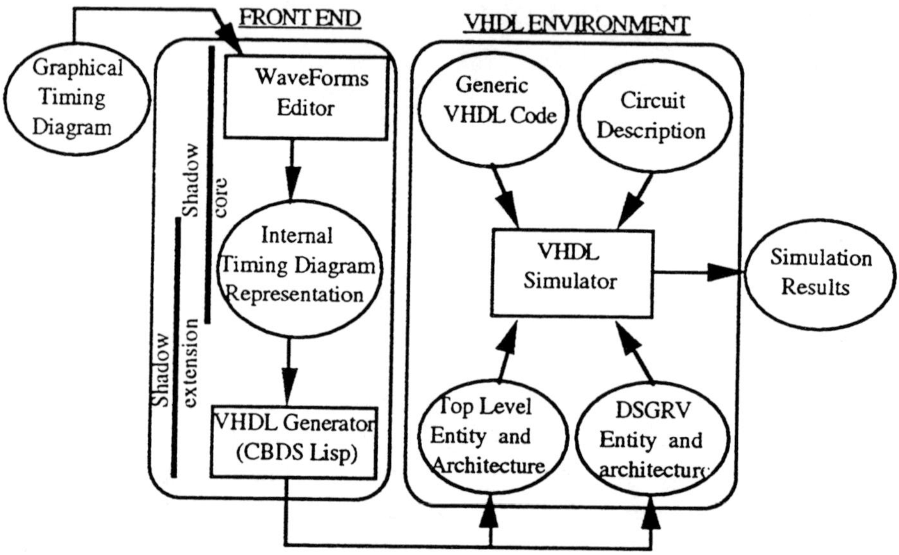

Fig. 1. Simulation System.

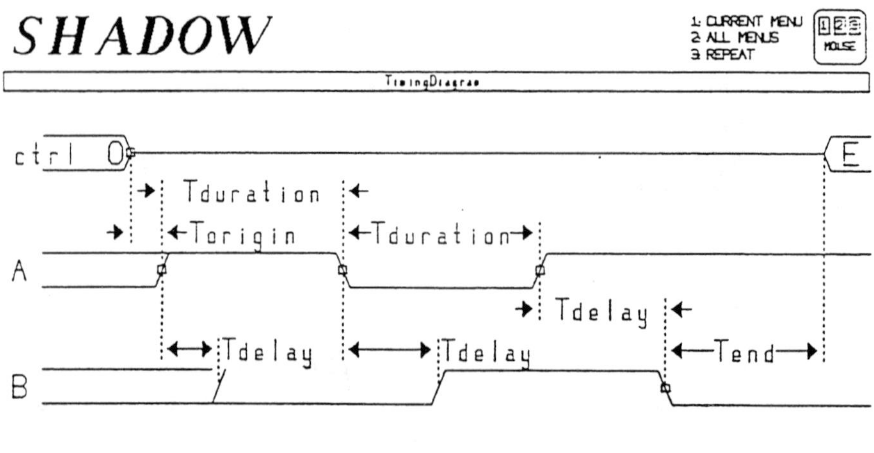

Fig. 2. Timing diagram representation with the waveform editor.

Hence, to simulate a circuit described in VHDL, we only need to specify, using the waveform editor, a composition of timing diagrams. Fig. 2 shows an example of a timing diagram produced using the waveforms editor.

3. The model

We use constraint graphs to represent timing diagrams. A constraint graph has a node representing each event in the timing diagram and directed edges between nodes representing timing constraints between events [2,3]. An edge of weight α from an event X to an event Y represents the inequality $T_y - T_x \geq \alpha$ where T_x (T_y) is the occurrence time of event X (Y). Two additional nodes, "origin" (O) and "end" (E), are inserted in the constraint graph. Event O represents time 0; an edge of weight zero is inserted from O to the first event of every signal. Event E can be used to fix the minimum and the maximum duration for the timing diagram. The set of constraints represented by the constraint graph is said to be consistent if there is a time assignment to the events such that all constraints are satisfied. It is well-known [3] that this property holds if and only if there are no positive cycles in the constraint graph.

We designate an event as "feasible" if it can be the next event to occur such that no constraints are violated. We can easily prove that an event Z is a feasible event if and only if all events X having a directed path with a weight greater than or equal to zero from X to Z, have occurred. When an event Z occurs at time T_Z we simply add an edge of weight T_Z from O to Z and an edge of weight $-T_Z$ from Z to O. We can then prove that the event Z must occur, to preserve consistency of the constraint system, in the time interval defined by [inf(Z),sup(Z)] such that inf(Z)=LP(O,Z), sup(Z)=-LP(Z,O), where LP(X,Y) indicates the weight of the longest directed path from X to Y. Fig. 3 shows the constraint graph representing the timing diagram of Fig. 2 with $T_{origin}=0$, $T_{duration}=3500$, $T_{end}=0$, $T_{delay(min)}=100$ and $T_{delay(max)}=500$. The origin node is considered as occurred as soon as the timing diagram begins to execute. If event #2 has occurred then event #5 is the only feasible event in the graph and we can compute the time interval for this event. An efficient algorithm to compute the feasible events set is given in [6].

4.1. VHDL PACKAGES

In this section we present the VHDL libraries developed for the DSGRV architecture. This VHDL code is "generic" in the sense that it is independent of the timing diagrams. Principal data structure definitions are shown in Fig. 4.

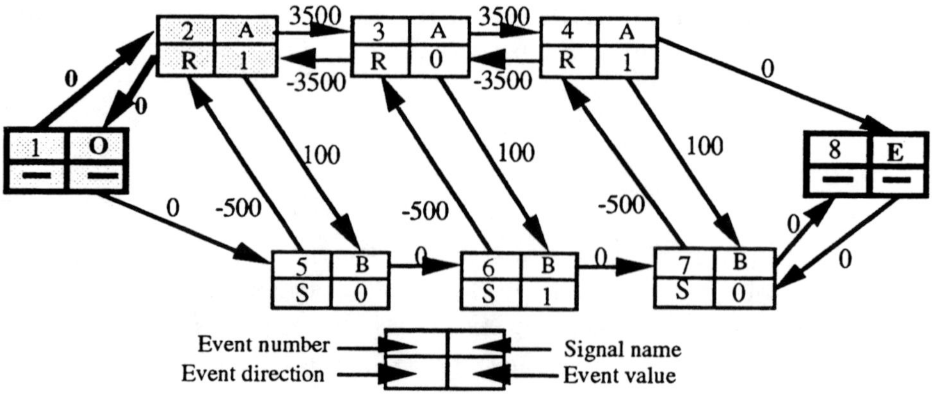

Fig. 3. Constraints graph.

```
package  data_structure is
type  sig_value  is  ('0','1','Z','X','V');
type  sig_direction  is  (stimulus,response);
type  graph_event;
type  graph_link  is  access  graph_event;
type  graph_event  is
      record
            event_label        : integer;
            signal_number      : integer;
            value              : sig_value;
            direction          : sig_direction;
                  :
      end record;
type  feasible_event;
type  feasible_event_set  is  access  feasible_event;
type  feasible_event  is
      record
            event              : graph_link;
            inf                : time;                 -- inf is LP(O,"event")
            sup                : time;                 -- sup is LP("event",O)
            next_feasible      : feasible_event_set;
      end record;
            :
end data_structure;
```

Fig. 4. Data structure.

The model uses the five value set {0,1,Z,X,S} where Z stands for high impedance, X means unknown or don't-care and S represents any arbitrary value that is stable for a specified period of time. Each event is represented by a graph node containing its direction, value, edges to other events, longest paths to other nodes and signal identifier. The feasible_event_set type is a list containing the next feasible events with their interval time computed by a longest path algorithm [5].

Two VHDL packages have been developed for the DSGRV architecture. The first one is the "Hierarchical_Utilities" package shown in Fig. 5. The procedures "Add_*" are used to build the structure representing the timing diagrams. The procedure "Initialization" initializes the timing diagram structure, i.e., the longest paths, the feasible events set, etc ... Two other procedures are used during the simulation run-time: "Find_feasible_events" returns the set of feasible events and "Update_Graph" updates the constraint graph. These are called at every event occurrence.

```
package hierarchical_utilities is
procedure add_dependencies(leaf : inout hierarchy; i,j : in integer;
                           weigth1,weigth2 : in time);
procedure add_concurrencies(leaf : inout hierarchy; i,j : in integer;
                           weigth : in time);
procedure add_son(h,son : inout hierarchy);
procedure initialisation(h : inout hierarchy);
procedure find_feasible_events(h : inout hierarchy;
                           f_events : inout feasible_event_set);
procedure update_graph(events : inout occurred_event_list;
                           empty : inout boolean);
end hierarchical_utilities;
```

Fig. 5. Hierarchical utilities package.

The other package developed for DSGRV is the "Simulation_Utilities" package (Fig. 6). This package contains procedures which are independent of the graph representation of timing diagrams. The procedures "Fix_Stimuli_Time" fixes an occurrence time for each stimulus event in the feasible set. There is an arbitrary choice of stimulus occurrence time in its interval. The procedure "Simulation_Max_Time" computes the maximum time until which the simulation is allowed to proceed without the occurrence of an event. This time must be the smallest of the minimum of stimuli event time and the minimum of sup(E_i), $E_i \in FS_R$ the set of feasible response events. This follows from the two following facts: First, simulation must stop if the time fixed for a stimulus event is reached; the event is then generated by the DSGRV with a zero delay. Second, simulation must stop if the maximum acceptable time for a response signal is reached because an event on this signal should have been observed; an error is then flagged. Finally, the procedure "Validate" validates all response events detected. To be valid, a response event must occur at a time included in its time interval and the value observed must match the value specified in the timing diagram.

```
package simulation_utilities is
procedure fix_stimuli_time(f_events : inout feasible_event_set;
                           min : inout time);
procedure simulation_max_time(f_events : inout feasible_event_set;
                           max_time : inout time);
procedure validate(f_events : inout feasible_event_set;
                           o_events_from_sim : inout events_to_verify_list;
                           o_event_to_update : inout occurred_event_list;
                           error : inout boolean; message : inout string(1 to 25));
end simulation_utilities;
```

Fig. 6. Simulation utilities package.

4.2. Generated VHDL Code

The two modules generated by the VHDL code generator are dependent on the timing diagrams specified by the user. These modules depend on signal direction, event values and timing constraints. The top level module contains the top level test bench entity and its architecture defining the ports mapping and a resolution function for bi-directional ports. The other module contains the DSGRV architecture, shown in Fig. 7. The procedure "Load_Structure" (Fig. 8) creates the constraint graph needed for the simulation.

```
entity ENTITY_NAME is  port( ...... );  end ENTITY_NAME;

architecture DSGRV of ENTITY_NAME is
begin
  process
     variable h              : hierarchy := null;                -- pointer on hierarchical root
     variable f_events   : feasible_event_set := null;              -- next feasible events
     variable events_from_sim : events_to_verify_list := null;  -- response events
     variable events_to_update: occurred_event_list := null;     -- list of occurred events
                                                                  --(stimuli and responses)
          :                                                  :
          :                                                  :
          procedure LOAD_STRUCTURE .....
               -- build the structure representing the timing diagram
          procedure GET_RESPONSES_TO_VERIFY .....
               -- return a list of response events that have actually occurred
          procedure GENERATE_STIMULUS .....
               -- generate stimulus with delay equal to zero
     begin
       :       -- main process controling simulation
  end process;
end DSGRV;
```

Fig. 7. DSGRV module.

```
procedure load_STRUCTURE(h : inout hierarchy) is
     variable access_table    : access_array;
     variable table           : lp_array;

begin
     access_table(1) := new graph_event'(1,0,'X',stimulus,true,0,null,null,table);
     access_table(2) := new graph_event'(2,1,'1',response,false,0,null,null,table);
     :
     h := new hierarchical_node'(1,leaf,null,null,0,0,0,0,0,8,access_table,
                                   access_table(1),access_table(8),null,null,null);

     add_dependencies(h,4,7,100,-500);
     add_dependencies(h,3,6,100,-500);
     :
end load_STRUCTURE;
```

Fig. 8. Load STRUCTURE procedure.

4.3. THE DSGRV PROCESS

The DSGRV is implemented as a single VHDL process (Fig. 9). The process validates responses and generates stimuli during the simulation run-time. The first block of instructions loads and initializes the constraint graph and it is done only the first time the process executes. The second block computes feasible events. It fixes a transition time for each stimulus in the feasible events set and computes the simulation timeout. Then, the DSGRV process waits until it wakes up due to the timeout or due to a transition on a response signal. Finally, any stimulus fixed at the wake-up time is generated, and the constraint graph is updated. The process stops when there are no more feasible events in the constraint graph or when an error occurs.

5. Hierarchical DSGRV

We have defined two timing diagram composition operations, TDConcat and TDConcur, which allow sequential and concurrent execution of timing diagrams, respectively. The possibility to compose timing diagrams in a hierarchical way (leaves are constraint graphs and timing diagrams are composed using "origin" and "end" nodes) keeps graph complexity and run-time at an acceptable level.

To adapt the DSGRV architecture to the hierarchical model, three procedures are modified: "Load_Structure", "Find_Feasible_Events" and "Update_Graph". "Load_Structure" now creates a hierarchical graph structure, "Find_Feasible_Events" percolates feasible events from the leaves to the root, and "Update_Graph" first updates the graph directly containing the occurred event; it then updates, in a bottom up fashion, all graphs that recursively contain that event. The interested reader should refer to [6] for additional details.

```
begin -- MAIN PROCESS controling simulation--
    if first_time then
        init_signals_state(signals_state);
        load_STRUCTURE(h);
        initialisation(h);
        first_time := false;
    end if;

    find_potential_events(h,p_events);
    fix_stimulus_times(p_events,actual_time);
    simulation_max_time(p_events,max_time);
    timeout := max_time - actual_time;
   wait on RESPONSE_SIGNALS for timeout;
    get_responses_to_verify(p_events,signals_state,events_from_sim);
    validate(p_events,events_from_sim,events_to_update,error,message,sig_number);
    if not error then
        actual_time := now;
        if actual_time = max_time then
            generate_stimulus(p_events,actual_time,events_to_update);
        end if;
        if events_to_update /= null then
            print_transitions(events_to_update);
            update_signals_state(events_to_update,signals_state);
            update_graph(events_to_update,empty);
        else
            error := true;
            message := "should have observed a transition";
    end if;  end if;
    if error or empty then
        if error then
            print_end(message,sig_number);
        else
            print_end("==> No error detected during simulation",-1);
        end if;
        stop;
    end if;
end process;
```

Fig. 9. DSGRV main process.

6. Experimental results

We have run different experiments using microprocessor bus operations. We present here READ transactions between the Multi-Protocol Communications Controller R68561 and a MC68000 processor. The experiments were performed on a SUN 3/260 running Intermetrics VHDL. Each chip was modeled as a separate DSGRV process using specifications of [7] for the

R68561 and [8] for the processor. The first experiment is a single READ operation using the "flat" model and the second is the concatenation of two READ operations using the hierarchical model. The number of nodes and edges used in each DSGRV process and the run-times are shown in Table 1. The term "updates" in this figure indicates the total number of times any one of the DSGRV processes was waken up due to an event activity.

A limitation encountered using the DSGRV system is the absence of conditionals in the semantics of the model. We have implemented TDChoice and TDLoop operations; however, these operations use static algorithms and are independent of circuit state.

Experiments	nb. of Nodes		nb. of Edges		run-time (sec.)	updates/ sec.
	MC68000	R68561	MC68000	R68561		
1 READ operation	24	16	104	66	4.9	7.4
2 READ operations (hierarchical)	50	34	214	138	10.0	7.2

Table 1. Simulation Results.

7. Conclusion

We have presented a new tool to perform stimulus generation and response validation using timing diagrams as a data source. The model is based on hierarchical timing constraint graphs. The DSGRV architecture dynamically interacts with the circuit during the simulation run-time; this interaction allows generation of stimuli in function of response events.

References

[1] G. Boriello. *A New Interface Specification Methodology and its Application to Transducer Synthesis.* PhD thesis, University of California, Berkeley, 1988.

[2] S.K. Sherman. *Algorithms for timing requirement analysis and generation.* In ACM/IEEE Proc. 25th DAC, pages 724-727, 1988.

[3] F. Jahanian and A.K.L. Mok. *A graph-theoritic approach for timing analysis and its implementation.* IEEE Transaction on computers, C-36(8), August 1987.

[4] VHDL Language Reference Manual, *IEEE Standard 1076-1987*, 1987.

[5] R.E. Tarjan *Data Structures and Network algorithms.* SIAM, Philadelphia, 1983.

[6] K.Khordoc, M.Dufresne and E. Cerny. *A stimulus/response system based on hierarchical timing diagrams.* Publication 770, Dept. I.R.O., Université de Montréal, 1991.
[7] Rockwell International.*Communication Products Data Book.* Rockwell International Corporation,1988.
[8] Motorola. *MC68000 16-bit microprocessor*. Motorola Inc., 1983.

Acknowledgments

The work was partially supported by NSERC Canada grant No MEF0040113, a BNR grant for VHDL modelling, and by the equipment loan of the Canadian Microelectronics Corporation.

Switch-Level Models in Multi-level VHDL Simulations

K. KHORDOC, M. BIOTTEAU, E.CERNY

Dép. d'IRO, Université de Montréal
C.P. 6128, Succ. A, Montréal, Qué., H3C 3J7 Canada

1. Introduction

Switch-level simulation is usually performed by special-purpose simulators [e.g., 3, 4, 7, 11, 17] which have a built-in knowledge of the semantics of switch-level modelling. However, in a modern digital system design environment such as VHDL [18], it is desirable to perform simulation at multiple abstraction levels, intermixing parts of the design described behaviorally at the specification level with other parts described at the gate or switch level. In this case the kernel of the simulator knows nothing about the semantics of each module, instead, the model of the module (e.g., transistor, capacitor, source) interacts with other such modules by the exchange of signals, and the simulator provides general mechanisms for synchronization and signal communication. It becomes thus necessary to capture the behavior of a transistor independently of its instantiation context (i.e., the transistor's state evolves strictly in reaction to the activity of signals on its ports), yet at the same time preserving the same overall behavior of the circuit as in a dedicated switch-level simulator where global knowledge of the (dynamically determined) channel connected component (CCC) is available. We call this approach *distributed switch-level modelling.*

Note that the proposed approach (similarly as in [9, 14, 15]) is appropriate only for multi-level VHDL simulation and cannot compete in performance with dedicated switch-level simulators such as [3, 4, 7], but the latter are unable to mix different levels of abstraction during the same simulation . To facilitate interfacing with higher abstraction levels, the type of the transistor gate port is the "logic4" type ('X', '0', '1', 'Z').

J. Mermet (ed.), VHDL for Simulation, Synthesis and Formal Proofs of Hardware, 43–62.

As indicated in Bryant's review paper [5], past attempts at including such switch-level models in gate-level simulators were far from satisfactory. As we show here, even more recent attempts [1, 14, 15, 16] do not completely achieve their goal, in part due to the lack of an adequate formal basis on which correct reasoning can be made. To avoid this problem we start with the MOSSIM-II algorithm [4] which is based on a solid formalism and modify it as follows: First we show that the MOSSIM-II algorithm can be carried out in one resolution pass instead of three to minimize event scheduling, and then we distribute the control of the one-pass algorithm to allow VHDL switch-level simulation. The signal reset and the bucket sort according to signal strengths in a CCC of a typical switch-level algorithm [e.g., 3, 4, 7] are achieved in a distributed manner using "delta" delay and 1 fs scheduling of events on port signals, respectively (i.e., in time using regular event scheduling mechanisms rather than in space using special sorting over a CCC).

In Section 2 we review recent distributed modelling approaches and identify their weaknesses. We show in Section 3 that the three evaluation phases of MOSSIM-II can be merged into one phase, and then adapt it for distributed VHDL modelling in Section 4. In Section 5 we present ways of improving the performance of our switch model. Experimental results are presented in Section 6 using the Intermetrics VHDL environment, and Section 7 concludes the presentation. The appendix contains the VHDL package and models.

2. Review of recent distributed switch-level algorithms

A signal is represented as an ordered pair of values <v, s>, where v is from the set of logic levels {L, H, X}, and s is from an ordered set of integers {0, ..., S_{max}} identifying the signal strength. Values $0 < s \leq S_{cap}$ represent storage node sizes, while values $S_{cap} < s < S_{max}$ represent transistor strengths (unless stated otherwise, the strength of a transistor represents the "magnitude" of its conductance in the ON state [8]). S_{max} corresponds to the strength of an input node (e.g., power supplies), while the signal <X, 0> represents the high-impedance state. We use $S_{cap} = 3$ and $S_{max} = 9$.

The first distributed algorithm was proposed by Kawai and Hayes [11], where a bidirectional transistor is modelled as a pair of unidirectional switches connected in parallel, but in opposite directions. When a conducting switch is processed, the value of the source is transmitted to the drain through one of the switches of the pair, and the drain value is transmitted to the source through the other switch. To avoid reflecting the value back to the switch (i.e., closing a loop within the pair of switches), an asymmetric connector (#) operation is used to exclude the contribution from the second switch of the pair. There is no reset to an initial value prior to evaluating the effects of a change in transistor setting, and the evaluation is not in a decreasing order of strengths.

When there is no cycle in the CCC, this approach is sufficient to account for the non-monotonic behavior of drivers and node strengths, and correct results are obtained. However, in the presence of cycles this method fails [5], because original stronger signal values can be memorized in the cycle when a change in transistor setting should lower the strength of the resulting signal. In fact, the problem is even more fundamental: it will fail even if a reset were used at the beginning of each evaluation, because path blocking [4] is not performed correctly. In Fig. 1 the problem would occur due to the memorization of <H, 5> in the loop N_2 - N_3, and the delayed arrival of <L, 8> at node N_1.

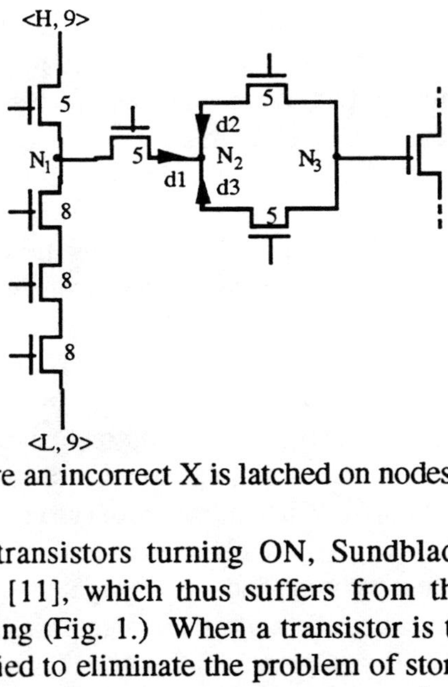

Fig. 1. Example where an incorrect X is latched on nodes N_2 and N_3.

In the case of transistors turning ON, Sundblad et al. [16] use an approach similar to [11], which thus suffers from the same problem of incorrect path blocking (Fig. 1.) When a transistor is turned OFF, a partial reset strategy is applied to eliminate the problem of stored values in loops. It consists of propagating <X, 0> values (pseudo-events) to all nodes affected by the transistor. This propagation is blocked at nodes where the driver carrying the pseudo-event is not the dominant (i.e., strongest) driver. After all pseudo-event activities terminate, the new dominator values are allowed to propagate their value to neighboring transistors. The reset strategy requires some control over the event scheduling queue of the simulator, which may not be possible in VHDL.

Stanculescu et al. [15] proposed the first switch-level model in VHDL, thus showing that it is possible to create models of bidirectional elements in this environment. A reset operation is performed on all nodes and transistors that are reachable from the event site through channel connections of transistors in the ON and X states. Consequently, the problem of stored values at the beginning of each new evaluation should not occur. The original algorithm has two problems, however, which were later corrected and the new algorithm is presented elsewhere in this volume. We identify these problems here in order to exemplify the type of difficulties that may be encountered when developing distributed switch-level models. First, the reset algorithm was incorrect, since the transistor which starts the reset does not wait (one delta) until the reset transaction propagates through the neighboring transistor(s), before commencing the actual evaluation. Hence, an incorrect dominating signal value can be computed in the originating transistor, as it may be based on a previous strong value from a neighbor. Second, the path blocking mechanism is incorrect: the transistor transfer function consists of the least upper bound (LUB) of the values on the transistor source and drain ports, limited (min.) by the transistor strength; the result is then propagated to both the source and the drain. There may thus be a "reflection" of the propagated value, resulting in X being latched in the network when the original driver changes value due to path blocking.

A potentially interesting solution is in [1, 2]: There is no reset at the beginning of evaluation, but a dynamically directional switch (DDS) is defined. A DDS drives one side at a time rather than propagating in both directions. The propagation is always from the stronger to the weaker side. It does not have the stored-value problem of Fig. 1, because the drivers d2 and d3 carry signals of strength 0 (high impedance). When d1 changes due to path blocking, the new value is correctly propagated in the subnetwork. Signal attenuation in transistors is done by the subtraction of transistor magnitude conductances, which can make the simulation results too optimistic without the Condfac safety factor of [16]. When a transistor turns OFF, it sets its drivers to <X, 0> (high impedance). If the driver was not a dominator on the node, the change does not affect the node and the propagation terminates there. If the driver were a dominator then the node value decreases either to a non-zero value, provided that there are other non-zero drivers on this node, or else to <X, 0> Subsequently, the neighboring transistors are evaluated, possibly causing change in the driving direction of some of them, etc.

In [1], the correctness of the DDS model is based on the proposition that there cannot be a directed cycle in the transistor graph, where a directed edge represents a DDS. However, this does not seem to be the case when an edge is deleted, as shown in the example of Fig. 2. Assume that all transistors are ON. The circuit settles in the state indicated in Fig. 2a (arrows show the driving direction of the transistors). Assume now that T_1 turns OFF. A signal <X, 0> is propagated from T_1, setting N_1 to <X, 0>. This is further propagated through T_2 and T_3, setting N_2 to <X, 0>. N_3 is dominated by <H, 5> from T_5 and thus N_3 is resolved to this value, as shown in Fig. 2b. Since N_3 is

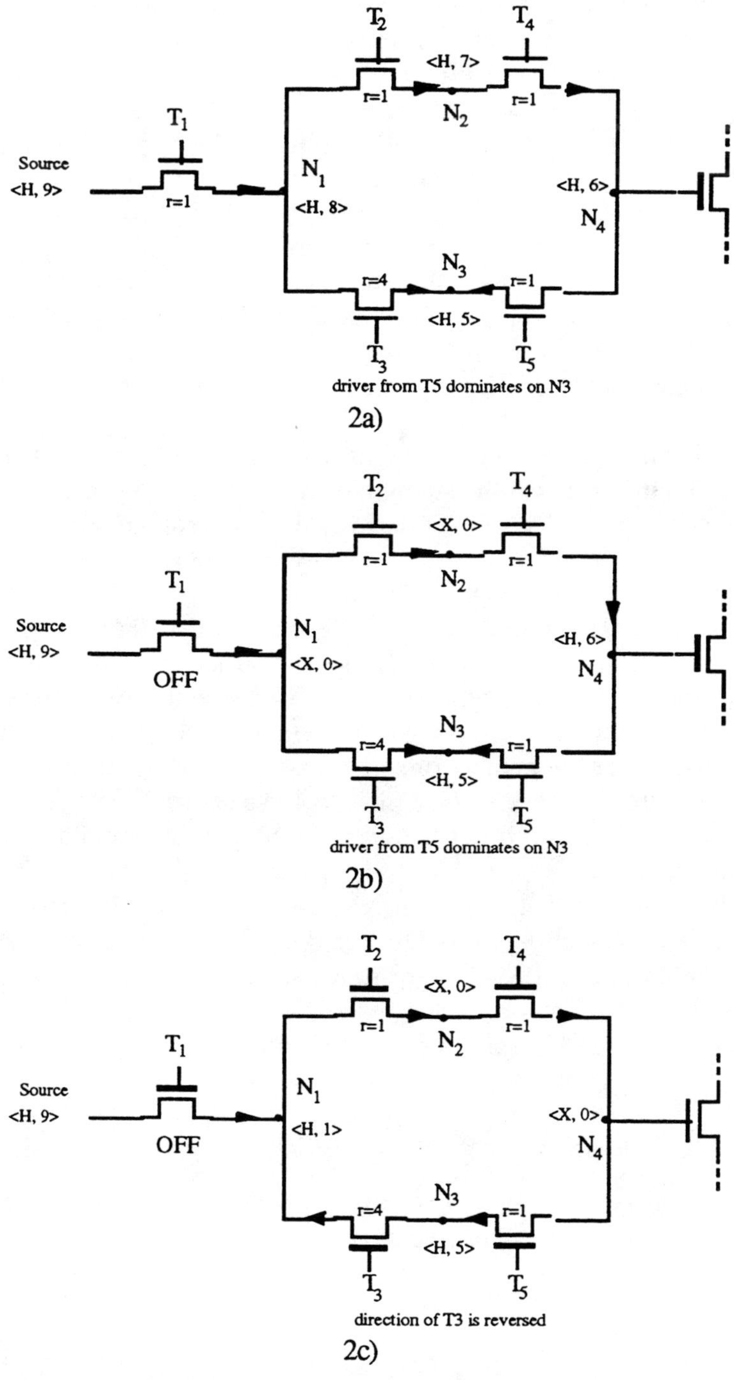

Fig. 2. Example where the method of [1] forms a directed cycle.

now stronger than N_1, the direction of T_3 is reversed and <H, 5> is propagated through T_3, resulting in <H, 1> on N_1. At the same time, <X, 0> on N_2 propagates through T_4, setting N_4 to <X, 0>, as shown in Fig. 2c. A **directed cycle N_1 -> N_2 -> N_4 -> N_3 -> N_1 has been created.** The cycle causes each node to oscillate between <X, 0> and <H, s>, s > 0, until s is attenuated to 0 by repeated passes through the cycle. The number of such wasted cycles can be considerable, because $S_{max} = 4096$ in [1]. Furthermore, such an oscillation would become persistent if subtraction were replaced by the min operation. A similar problem may be encountered in [14]. In [2], a solution to this problem is introduced: all signals of driven strength ($s > S_{cap}$) are transmitted only after all possible transmissions of capacitive strength ($s \leq S_{cap}$) signals are performed. This solution, however, is incompatible with a totally distributed approach, because it requires control of the scheduling queue.

The principal reason for the difficulties in the above methods is that the condition under which the strength of a signal is allowed to decrease is not handled properly. In MOSSIM-II [4] this is resolved using a reset at the beginning of each evaluation, and by carrying out the evaluation in a decreasing order of strength values.

In the next section we show that the MOSSIM-II algorithm can be implemented using *one resolution phase* rather than three as in [4]. The motivation is to synchronize the evaluations, and to minimize the average number of transistor model evaluations and generated events. This is important in VHDL, since the number of source and drain events directly influences the overall simulation performance (note that in a dedicated switch-level simulator such an optimization is not as important, because it would only save the loop control overhead).

We adapt in Section 4 the resulting one-pass algorithm to distributed switch-level modelling in VHDL. We thus can capitalize on the sound formal basis of the MOSSIM-II algorithm and, as a result, avoid the problems outlined in this section. In the remainder of the paper, we call the original MOSSIM-II algorithm as "three-phase" and our version as "one-phase".

3. One-phase MOSSIM-II algorithm

We present the one-phase algorithm in pseudo-code and then show that it is equivalent to the three-phase version.

3.1 EXPLANATORY NOTES FOR THE PSEUDO-CODE

- Strength representation: strength (magnitude conductances) are positive integers from 0 to S_{max} (Section 2).
- *nodes_to_solve* is the set of nodes on which state values must be recomputed due to a gate event. We assume that this set ("vicinity" in [4]) is computed as in the three-phase algorithm.
- As in the three-phase algorithm we use a bucket sort array to control the evaluation in a decreasing order of strengths. In our case the values inserted in the array are triplets (S_{block}, S_{low}, S_{high}), the components of which are defined as in MOSSIM-II. The three values S_{block}, S_{low} and S_{high} are computed in a single sweep of the bucket array.
- An event E in the bucket array is a pair (Node, Value), where Node is from nodes_to_solve and Value is a triplet (S_{block}, S_{low}, S_{high}).
- *min_non_zero(Value)* returns the smallest non-zero element of a triplet. It is used to determine the bucket into which an event is to be inserted (see Lemma 1 below).
- *event_node(E)* and *event_value(E)* are accessors for Node and Value of E, respectively, where E is an event.
- *make_event(N, Val)* constructs an event for node N and value Val.
- *insert_into_bucket(S, E)* inserts event E into the bucket of strength S.
- *bucket(S)* returns the list of events in the bucket of strength S.
- *node_value(N)* is the accessor for the value (i.e., triplet) of node N.
- *value_block(Value)*, *value_low(Value)* and *value_high(Value)* are the respective accessors for S_{block}, S_{low} and S_{high} in Value.
- *transistor_other_node(Transistor, fromNode)* returns the source (resp. drain) of Transistor given its drain (resp. source).
- *transistor_strength(Transistor)* returns the strength of Transistor.
- *transistor_state(Transistor)* returns the state of Transistor (one of {ON, OFF, X}).
- *source_value(Node)* returns the value of the source (driving or capacitive) connected to Node. This value can be (S,S,0), (S,0,S) or (S,S,S), S > 0.

3.2 PSEUDO-CODE OF THE ONE-PHASE ALGORITHM

When a change in the logic value on a transistor gate is detected, the set nodes_to_solve is computed as in MOSSIM-II and the following procedure RESOLUTION_CYCLE is executed:

```
RESOLUTION_CYCLE( nodes_to_solve ):
    INITIALIZE( nodes_to_solve );
    RESOLVE ();
```

```
    Compute logic values on affected nodes from the triplets as in MOSSIM-II;
End RESOLUTION_CYCLE.

INITIALIZE( nodes_to_solve ):
 For each N in nodes_to_solve do
  node_value(N) := (0,0,0);            /* initialize node */
  value := source_value( N );
  insert_into_bucket( value_block( value ), make_event( N, value ) );
                                                   /* init. bucket array */
 End For;
End INITIALIZE.

RESOLVE( ):
 For S from S_max down to 1 do
  For each event E in bucket(S) do
   remove E from bucket(S);
   N := event_node(E);
   previous_node_value := node_value(N);
   node_value(N) := node_resolution(N, event_value(E));
   If node_value(N) /= previous_node_value     /*any triplet element changed*/
    then
    For each transistor T whose src or drn is connected to N do
      transistor_propagate(T, N); end For;
   end If;
  end For;
 end For;
End RESOLVE.

NODE_RESOLUTION( node, newVal ):
 nodeVal := node_value(node);
 /* determine the Least Upper Bound (LUB) of nodeVal and newVal: */
 S_block := max( value_block(newVal), value_block(nodeVal) );
 S_low   := max( value_low(newVal),   value_low(nodeVal) );
 S_high  := max( value_high(newVal),  value_high(nodeVal) );
 If S_low  < S_block then S_low  := 0 end If;           /* path blocking: */
 If S_high < S_block then S_high := 0 end If;
 return( make_triplet( S_block, S_low, S_high ) );
End NODE_RESOLUTION.

TRANSISTOR_PROPAGATE(transistor, fromNode):
 state := transistor_state(transistor);
 If state /= OFF
  then
  strength := transistor_strength(transistor);
  new_value := transistor_transfer(fromNode, strength, state);
  otherNode := transistor_other_node(transistor, fromNode);
  b_strength := min_non_zero( new_value );   /* bucket strength */
  insert_into_bucket(b_strength, make_event(otherNode, new_value));
 end If;
```

```
End TRANSISTOR_PROPAGATE.

TRANSISTOR_TRANSFER(fromNode, transistor_strength, transistor_state):
  out_value := nil;
  Loop for i from 1 to 3
      for S in node_value(fromNode)
      for effective_state in effective_transistor_state(transistor_state)
        do
          If effective_state = ON
           then
             out_value(i) := min(S, transistor_strength)
           else
             sout_value(i) := 0
          end If;
  end Loop;
  return(out_value);
End TRANSISTOR_TRANSFER.

EFFECTIVE_TRANSISTOR_STATE(transistor_state):
  Case transistor_state
    ON:     return (ON, ON, ON);
    X:      return (OFF, ON, ON);
    OFF:    return (OFF, OFF, OFF);
  end Case;
End EFFECTIVE_TRANSISTOR_STATE.
```

3.3 Equivalence with Three-Phase MOSSIM-II

Lemma 1: An event of value $V=(S_{block}, S_{low}, S_{high})$ must be inserted into the bucket with index S = min_non_zero(V), i.e., the smallest non-zero component of V.

Proof: Since we process the buckets in a decreasing order of strengths, any component of V greater than S had to be propagated earlier. Therefore, the component having the smallest non-zero value caused the event. It cannot be 0 because such strength indicates no propagation yet. Q.E.D.

The following two lemmas and Theorem 1 establish the equivalence of the one- and three-phase algorithms.

Lemma 2: The computed value of S_{block} on any node is the same in the one-phase and the three-phase algorithms.

Proof: S_{block} is independent of S_{low} and S_{high}, and it is computed as in the three-phase algorithm (reset to 0, propagate in a best-first manner [13], and update by the max(current node value, event's S_{block} value)). Q.E.D.

Lemma 3: The computed S_{low} and S_{high} on any node is the same in the one-phase and the three-phase algorithms.
Proof: (We prove the lemma for S_{low}, it is similar for S_{high}.) S_{low} is influenced by S_{block} only when $S_{block} > S_{low}$. Since evaluation is performed in a decreasing order of strengths, such an S_{block} value arrives to the node strictly before the weaker S_{low}. Therefore, S_{low} is the same as in the three-phase case where the S_{block} is evaluated before S_{low}. Q.E.D.

Theorem 1: The one-phase algorithm is equivalent to the three-phase version (i.e., both yield the same triplet of values on all nodes).
Proof: Follows from Lemmas 2 and 3.

The following definition and lemmas will be used to establish the correctness of the VHDL implementation of the one-phase algorithm (Sect.4).

Definition 1: A component of a node value triplet is one-step non-decreasing if throughout the node value evaluation in a CCC it either remains at 0, or changes once from 0 to some $S > 0$ and then remains unchanged till the end of the evaluation.

Lemma 4: The value of S_{block} on any node is one-step non-decreasing.
Proof: The node's S_{block} value is initialized to 0 and then updated by taking the max of its current value and the event's $S_{block.e}$ value. Hence, S_{block} cannot decrease. Due to the order of evaluation, the highest possible value that S_{block} can take is assigned first. Q.E.D.

Lemma 5: The value of S_{low} (resp. S_{high}) on any node is one-step non-decreasing.
Proof: (We prove the lemma for S_{low}, the proof is similar for S_{high}). Since S_{low} is independent of S_{high}, we ignore the S_{high} component here for reasons of simplicity. Node values thus consist of pairs (S_{block}, S_{low}). Only the following four valid node values can occur:

Case 1:	(0,0)	
Case 2:	(S,0),	$S > 0$
Case 3:	(0,S),	$S > 0$
Case 4:	(S_1,S_2),	$0 < S_1 <= S_2$

The case (S_1, S_2), $S_1 > S_2 > 0$ cannot be produced by the NODE_RESOLUTION procedure. Similarly, a valid event value falls under Case 2, 3 or 4. By exhaustive analysis of the valid node and event value triplets, it can be shown that either S_{low} remains unchanged or it increases only when it was 0, and that the resulting node value is one of the four valid cases (recursive argument), and thus S_{low} is one-step non-decreasing. Q.E.D.

It can be easily shown that the number of times the model of a transistor is evaluated (i.e., execution of the TRANSISTOR_PROPAGATE procedure) in the one-phase algorithm is smaller or equal to the number of such evaluations in the three-phase algorithm.

4. Distributed switch-level modelling in VHDL

In this section the one-phase algorithm of Sec. 3 is modified to yield a distributed control VHDL implementation.

Our VHDL model of a switch-level circuit consists of an interconnection of the instances of transistor, input source and capacitance entities (their VHDL code is in the Appendix). The transistor has three ports: the gate is an input port, and the source and drain ports are bidirectional. The input source has a single bidirectional port which is used in input mode for waking up the source (to inject its value at the appropriate time), and in output mode for actually injecting the source value. The capacitance has two ports: the first one is bidirectional and plays a similar role as the port of the input source; the second port is an output through which the capacitance drives its fan-out transistor gates.

Model execution (simulation) consists of a repetition of CCC resolution cycles. Each cycle has a time duration of cycle_time = (smax + 1) * d, where smax is the maximum strength, as defined in Section 2, and d is some small time constant (we use smax = 9 and d = 1fs). A CCC cycle is functionally equivalent to the RESOLUTION_CYCLE procedure of Sec. 3, except that the ordered bucket array sweep is emulated by the propagation of sources and signals of a given strength at precise time instants during the cycle, as explained below.

A CCC resolution cycle is triggered by the occurrence of an event at the gate of a transistor T; assume that such an event occurs at a time $t = t_0$. The transistor then resets its source and drain ports by driving them to (S_{block}, S_{low}, S_{high}) = (0, 0, 0) with one delta delay (one VHDL execution cycle, no advancement in simulation time). The resulting transactions on the drain and source signals of transistor T activate all neighboring entities (transistors, capacitances and sources) at the next delta cycle. Each such entity performs the same reset action (except for transistors that are OFF: they had their drain and source drivers already set to (0,0,0) the last time they participated in a subnetwork resolution, and do not contribute anything in the current resolution cycle). Furthermore, each source or capacitance that was activated schedules an event corresponding to its source value at a future time $t_s = t_0 + [(smax - s + 1) * d]$, where s is the source strength (driving or capacitive). The "reset" operation propagates recursively using delta delays (in zero simulation time) to all subnetwork elements that are reachable through channel connections of ON and X state transistors from the source and drain nodes of transistor T. Note that the source value is constant from one cycle to the next, while in the case of a capacitance it is the value memorized from the previous resolution cycle involving that capacitance. Also, to avoid proliferation of useless activity, every transistor driven by a source must have its own instance of the input source entity connected to the appropriate (source or drain) port.

After performing its reset and source value scheduling, each activated source and capacitance suspends itself for the duration of cycle_time, and then becomes ready for a new resolution cycle. On the other hand, each activated transistor must stay awake to perform subnetwork resolution activities, i.e., the propagation of source values. However, prior to this and right after performing the reset, each such transistor suspends itself for three delta cycles (using three "wait for 0 fs" statements), in order to avoid the reactivation of the same transistor at the next delta cycle due to transactions caused by the reset. The reset operation has the combined effect of computing nodes_to_solve and performing the INITIALIZE procedure of Section 3.

When simulation time advances to t_s such that $t_0 + d \leq t_s \leq t_0 + smax * d$, the scheduled source events of strength s take effect. Furthermore, all signals of strength s are propagated in the subnetwork (by this time, the transistors which were put to sleep for three delta cycles are already awake and ready for the signal propagation activity). Transistors propagate events with a delay of max(0, (event_strength - transistor_strength)) * d, where event_strength is the min_non_zero strength in the event triplet. This delay emulates the scheduling of events in the appropriate buckets.

Finally, when simulation time advances to $t = t_0 +$ cycle_time, the subnetwork is stable, since all signals (of strengths smax down to 1) have been propagated. All capacitances that were involved in the network resolution cycle now read the triplet value that settled at their node, and drive their fan-out transistor gates with the corresponding logic state using some pre-assigned inertial delay delay_to_gate. In order to achieve correct synchronization, all entities (transistors, input sources and capacitances) must be supplied with the same values of the generic parameters smax and d. Furthermore, delay_to_gate should be orders of magnitude greater than d * smax, and, to assure that the resolution cycle is never interrupted by gate events, delay_to_gate should be divisible by (smax + 1) * d. For instance, d = 1fs, smax = 9, and delay_to_gate in ns or ps are practical for MOS circuits. Alternately, this problem can also be solved using a filter process on transistor gates which assumes that a new evaluation cycle is not started before completing the current one.

A VHDL signal resolution function is provided for node signals (function resolution_triplet). This is a simple max operation combined with path blocking as in NODE_RESOLUTION of Sect. 3, except that the node value is always computed from all the contributing drivers, rather than cumulatively as in the one-phase algorithm. These resolution methods are equivalent, due to the non-decreasing property of node values (Lemmas 4 and 5).

5. Algorithmic improvements

We present three performance improving modifications to the model of Sect. 4. The first modification is trivial and consists of making the transistor propagate a value through its channel only if this value modifies the signal value at the destination port. This reduces the amount of simulation activity.

The second modification uses a signal representation consisting of pairs of values (Vl, Vh) rather than triplets, where Vl (resp. Vh) represents the strongest possible contribution of Low and X sources (resp. High and X sources) 10]. Each of Vl and Vh encodes both state and strength into one symbol, using an algebra based on an enumerated set. The transistor model is modified to transmit Vl (resp. Vh) if the transistor is either in the ON state or else if it is in the X state and the state of Vl (resp. Vh) is Low or X (resp. High or X). The node resolution function computes independently two least upper bounds, one for Vl and one for Vh. At the end of the resolution cycle, the final node state is set to the common state of its Vl and Vh components if they are the same, otherwise it is set to X.

The third modification consists of scheduling the injection of all capacitive sources at the same time step in the resolution cycle. This reduces the simulation time span of the resolution cycle. Correctness of the model is preserved because transistors do not attenuate capacitive values, and thus the order in which the values are propagated does not affect the final node values.

6. Experimental results

We illustrate the run-time performance of our switch-level model on benchmarks and show its usefulness in multi-level simulation. Capacitances are placed only at nodes driving transistor gates. The interface between a switch-level subcircuit and one modelled at a higher level is unidirectional (logic4 type).The experiments were done using the Intermetrics VHDL simulator v2.1, on a SUN 3/260 with OS 4.0.3.

In Table 1, we compare the run-time performance of the switch-level model with other levels of representation for a 16-bit ripple carry adder.

GL_ADDER16 and CCG_ADDER16 are respectively a gate-level and a switch-level CMOS complex-gate version of the adder (Fig. 8.1 (a) and (b) in [19]). EG_ADDER16 is derived from GL_ADDER16 by expanding each gate to the switch level (note that the number of gates increases due to the addition of inverters in the expansion of the non-inverting gates). BL_ADDER16 is a behavioral version at the bit level in which each one-bit adder is described as two concurrent signal assignments (sum and carry). The model used for the switch-level versions (CCG and EG adders) includes both the selective signal propagation and the modified signal representation outlined in Section 5. The stimuli are always the same set of 512 randomly generated pairs of 16-bit numbers. Notice that there is a difference of an order of magnitude in performance between the gate-level (GL_ADDER16) and the switch-level (CCG_ADDER16) versions.

Circuit	CPU Time	# of transactions
CCG_ADDER16	14 min 51 s	1 825 634
EG_ADDER16	29 min 23 s	2 720 512
GL_ADDER16	01 min 15 s	32 503
BL_ADDER16	51 s	20 500

Table 1. Simulation results for different models of ADDER16.

In Table 2, we compare the run-time performance of the first three versions of the switch-level model. The first version is the model of Section 4, the second version contains the selective signal propagation and the third version includes both the selective signal propagation and the Vl/Vh signal representation. The first two circuits are EG_ADDER16 and CCG_ADDER16 as described above. MC_ADDER16 is a 16-bit precharged Manchester carry chain adder (Fig 8.14 of [19]), in which the G and P signal generators are modeled at the gate-level, while the carry chain is at the switch level. The input sequence is the same as in the experiments of Table 1. BARREL16 is a 16-bit barrel shifter (Fig 5.15 of [12]) for which the stimuli consist of sixteen shift operations (shift by 0, shift by 1, ... until shift by 15) of a 16-bit word. RAM4X4 is a 16-bit RAM similar to the one described in [6]. The memory cells are modelled at the switch-level and are composed of two cross-coupled inverters and two transmission gates. The bit lines are driven by a bidirectional mux/demux modeled at the switch level. It is composed of a tree of transmission gates, controlled by the lower half of the address bus. The word lines which control the transmission gates of the memory cells are generated using a decoder modeled as a single entity at the behavioral level. The decoder is fed by the upper half of the address bus. The address bus is latched using DOMINO-type gates modelled at the switch level. The stimuli consist of writing an alternating sequence of 0's and 1's into the sixteen consecutive memory cells, followed by sixteen READ operations using the same address sequence.

The selective signal propagation improves CPU time by 4% (RAM4X4) to 11% (CCG_ADDER16) with respect to the original model. When the Vl/Vh algorithm is used with selective signal propagation, the improvement ranges from 21% (BARREL16) to 27% (MC_ADDER16).

Circuit	# of transist.	Original	Selective propagation	Vl / Vh
EG_ADDER16	832	38 min 09 s 4 062 954	34 min 30 s 3 625 355	29 min 23 s 2 720 512
CCG_ADDER16	448	20 min 21 s 2 853 025	18 min 05 s 2 409 024	14 min 51 s 1 825 634
MC_ADDER16	69	22 min 19 s 4 384 273	21 min 13 s 3 903 288	16 min 35 s 2 906 009
BARREL16	288	29 s 97 698	27 s 86 434	23 s 70 722
RAM4X4	210	5 min 33 s 1 025 057	5 min 21 s 911 960	4 min 07 s 678 051

Table 2: Performance of different versions of the evaluation algorithm (CPU time and the number of transactions)

Circuit	Vl/Vh: 1fs for capacitors	Vl/Vh: delta for capacitors	number of capacitances
EG_ADDER16	35 min 49 s 3 035 135	32 min 59 s 3 035 135	256
CCG_ADDER16	21 min 25 s 2 555 631	20 mn 30 s 2 555 631	304
MC_ADDER16	26 min 51 s 4 742 233	26 mn 27 s 4 600 449	181
BARREL16	26s 80 850	26s 80 850	96
RAM4X4	4 min 43s 768 163	4 mn 39 s 761 418	100

Table 3: Performance of the last two versions of the algorithm after adding capacitances of different sizes

The circuits used in the experiments in Tables 1 and 2 had few capacitances of the same size (3). In order to test the improvements brought by the transmission of capacitive values in delta-delay, we added several capacitances of various sizes to the circuits and applied the same input stimuli as before. Two versions of the Vl/Vh algorithm were used, one with scheduling of capacitive values at 1fs and the other with propagation using delta delays. The results are in Table 3: the propagation of the capacitive values in delta delays improves CPU time by 0% (BARREL16) to 8% (EG_ADDER164X4). The absence of visible improvement for the BARREL16 is due to the short simulation time and a small number of capacitances in the circuit.

7. Conclusions

We analyzed the weaknesses of existing distributed switch-level algorithms, and proposed a provably equivalent distributed version of the MOSSIM-II algorithm. Its implementation in VHDL was described, where transistors, input sources and capacitances are independent entities. No global control signals are required. This implementation is not as efficient as a dedicated switch-level simulator; but it is suitable for multi-level simulations. The gate signal type is the commonly used 4 valued logic ('X', '0', '1', 'Z'). Experimental results show that for circuits completely described at the switch level the performance is about 10x lower than with gate level models.

References

[1] D. Adler, "A Dynamically-Directed Switch Model for MOS Logic Simulation," 25th Design Automation. Conf., 1988, pp.506-511.

[2] D. Adler, "Switch-level Simulation Using Dynamic Graph Algorithms", IEEE transactions on CAD of integrated circuits and systems, Mar. 1991, pp. 346-355.

[3] P. Agrawal, S.C. Robinson, T.G. Szymanski, "Automatic Modeling of Switch-Level Networks Using Partial Orders," IEEE TCAD,Vol. 9, No. 7, July 1990, pp.696-707.

[4] R.E. Bryant, "A Switch-Level Model and Simulator for MOS Digital Systems," IEEE TC, Feb. 1984, pp.160-177.

[5] R.E. Bryant, "A Survey of Switch-Level Algorithms," IEEE Design & Test Magazine, Aug.1987, pp.26-40.

[6] R.E. Bryant, "Formal Verification of Memory Circuits by Switch-Level Simulation", Report CMU - CS - 89 - 156, June 1989.

[7] R.H. Byrd, G.D. Hachtel, M.R. Lightner, M.H. Heydemann, "Switch-Level Simulation: Models, Theory and Algorithms," in *Advances in Computer Aided Engineering Design*, (A. Sangiovanni-Vincentelli, Editor), Jai Press Inc., 1985.

[8] E. Cerny, J.P. Hayes, N.C. Rumin, "Magnitude Classes in Switch-Level Modelling," ICCD'89, Sept.89, pp.284-288.

[9] D. Coelho, "A VHDL Standard Package for Logic Modelling," IEEE Design & Test of Computers, June 1990, pp.25-32.
[10] J. Gecsei, E. Cerny, "Self-Adjusting Networks for VLSI Simulation," IEEE TC, Sept.1987, pp.1114-1120.
[11] M. Kawai, J.P. Hayes, "An Experimental MOS Fault Simulation Program CSASIM," Proc. Design Automation Conf., 1984, pp.2-9.
[12] C. Mead, L. Conway, Introduction to VLSI Systems, Addison-Wesley, 1980.
[13] R. Sedgewick, *Algorithms*, Computer Science Series, Addison-Wesley, 1983.
[14] S.P. Smith, R.D. Acosta, "A Value System for Switch-Level Modelling," IEEE Design & Test of Computers, June 1990, pp.33-41.
[15] A.G. Stanculescu, A.S. Tsay, A.N.D. Zamfirescu, D.L. Perry, "Switch-Level VHDL Descriptions," ICCAD'89, 1989, pp.180-183.
[16] R. Sundblad, C. Svensson, "Fully Dynamic Switch-Level Simulation of CMOS Circuits," IEEE TCAD, Mar.1987, pp.282-289.
[17] C.J. Terman, *Simulation Tools for Digital LSI Design*, Ph.D. Dissertation, M.I.T., Sept.1983.
[18] *IEEE Standard VHDL Language Reference Manual,* IEEE Std 1076-1987, 1988.
[19] N. Weste, K. Eshraghian, Principles of CMOS VLSI Design, Addison-Wesley, 1985.

Acknowledgments

The work was partially supported by NSERC Canada grant No MEF0040113, a BNR grant for VHDL modelling, and by the equipment loan of the Canadian Microelectronics Corporation.

Appendix

```
-- VHDL package and entities for switch-level modelling
-- (using one phase implementation of MOSSIM-II)

package switch_pack is
  constant smax: integer := 10;
  type strength is range 0 to smax; type triplet is array(1 to 3) of strength;
  type triplet_vector is array(natural range <>) of triplet;
   function resolution_triplet(tri_v: triplet_vector) return triplet;
   subtype resolved_triplet is resolution_triplet triplet;
  type logic4 is ('X','0','1','Z'); type logic4_array is array(natural range <>) of logic4;
  type triplet_array is array(natural range <>) of resolved_triplet;
   type polarity is (n,p); type gate_state is (on_state,off_state,unknown_state);
   type activity is (active,inactive);
  constant d: time := 1 fs; constant cycle_time: time := (smax+1)*d;
   function min_non_null(input1,input2,input3: strength) return strength;
   function max(input1,input2: integer) return integer;
    function transistor_state(transistor_polarity: polarity;
                              gate: logic4) return gate_state;
   function transistor_transfer(terminal: triplet; tr_strength: strength;
                              sstate: gate_state) return triplet;
```

```
  function voltagestrength_to_triplet(source_voltage: logic4;
                              source_strength: strength) return triplet;
  function attenuate_triplet(io_port: triplet; size:strength) return triplet;
  function triplet_to_voltage(io_port: triplet) return logic4;
  procedure propagate_after_delay(signal sending_port: in resolved_triplet;
                    signal receiving_port: out resolved_triplet;
                    transistor_strength: in strength;  sstate: in gate_state);
end  switch_pack;

package  body  switch_pack  is
  function resolution_triplet(tri_v: triplet_vector) return triplet is
    variable result_triplet: triplet;
  begin
   for i in 1 to 3 loop result_triplet(i) := 0; end loop;
   for i in tri_v'range loop  for j in 1 to 3 loop
      if tri_v(i)(j) > result_triplet(j) then result_triplet(j) := tri_v(i)(j); end if;
    end loop; end loop;
   if result_triplet(2) < result_triplet(1) then result_triplet(2) := 0;
     elsif result_triplet(3) < result_triplet(1) then result_triplet(3) := 0;  end if;
   return result_triplet;
  end resolution_triplet;

  function min_non_null(input1,input2,input3: strength) return strength is
   variable result: strength;
  begin
   if input1 /= 0 then result := input1; elsif input2 /= 0 then result := input2;
                                             else result := input3;  end if;
   if input1 < result and input1 /= 0 then result := input1; end if;
   if input2 < result and input2 /= 0 then result := input2; end if;
   if input3 < result and input3 /= 0 then result := input3; end if;
  return result;
 end min_non_null;

 function max(input1,input2: integer) return integer is
 begin
  if input1 > input2 then  return input1; else return input2; end if;
 end max;

 function transistor_state (transistor_polarity: polarity; gate: logic4)  return gate_state is
 begin
  if gate = 'X' or gate = 'Z' then return unknown_state;
    elsif transistor_polarity = n then if gate = '1' then return on_state;
                                           else  return off_state;  end if;
    elsif gate = '0' then return on_state; else return off_state;  end if;
 end transistor_state;

 function transistor_transfer(terminal: triplet;tr_strength: strength;
                                 sstate: gate_state) return triplet is
  variable result: triplet;
   function min(input1,input2: strength) return strength is
  begin  if input1 < input2 then return input1; else return input2; end if;  end min;
 begin
  for i in 1 to 3 loop result(i) := min(terminal(i),tr_strength); end loop;
  if sstate = unknown_state then result(1) := 0; end if;
```

```
    return result;
  end transistor_transfer;

   function voltagestrength_to_triplet(source_voltage: logic4;
                                       source_strength: strength) return triplet is
    variable result_triplet: triplet;
  begin
    result_triplet := (0,0,0);
    if source_voltage = '0' then result_triplet(1) := source_strength;
                              result_triplet(2) := source_strength;
      elsif source_voltage = '1' then   result_triplet(1) := source_strength;
                                        result_triplet(3) := source_strength;
      else result_triplet(2) := source_strength;  result_triplet(3) := source_strength; end if;
    return result_triplet;
   end voltagestrength_to_triplet;

   function attenuate_triplet(io_port: triplet;size: strength) return triplet is
    variable result_triplet: triplet;
  begin
    result_triplet := (0,0,0);
    for i in 1 to 3 loop if io_port(i) > 0 then result_triplet(i) := size; end if;  end loop;
     return result_triplet;
  end attenuate_triplet;

   function triplet_to_voltage(io_port: triplet) return logic4 is
  begin
    if io_port(1) > io_port(2) then return '1';
                    elsif io_port(1) > io_port(3) then return '0';  else return 'X'; end if;
   end triplet_to_voltage;

  procedure propagate_after_delay(signal sending_port: in resolved_triplet;
      signal receiving_port: out resolved_triplet; transistor_strength: in strength;
      sstate: in gate_state) is
    variable present_strength: strength;  variable delay: time;
  begin
   present_strength :=
                min_non_null(sending_port(1),sending_port(2),  sending_port(3));
   delay :=
       (max(0,strength'pos(present_strength)-strength'pos(transistor_strength)))*d;
    receiving_port <=  transport transistor_transfer(sending_port, transistor_strength,
                                                              sstate) after delay;
  end propagate_after_delay;
 end  switch_pack;

 entity  transistor  is
   generic(transistor_polarity:  polarity;transistor_strength:  strength);
  port(source,drain: inout resolved_triplet;gate: in logic4 := 'X');
 end  transistor;

 architecture  switch  of  transistor  is
  signal mode: activity := inactive;
 begin
  process
    variable sstate: gate_state; variable present_strength: strength; variable delay: time;
```

```
  begin
    sstate := transistor_state(transistor_polarity,gate);
   if mode = inactive  then  if gate'event or sstate /= off_state then
          source <= (0,0,0); drain <= (0,0,0);
          mode <= active after d, inactive after cycle_time;
          wait for 0fs; wait for 0fs;wait for 0fs; end if;
     elsif source'event or drain'event then if sstate /= off_state then
           if source'event then
             propagate_after_delay(source,drain,transistor_strength,sstate); end if;
     if drain'event then
           propagate_after_delay(drain,source,transistor_strength,sstate); end if;
  end if; end if;
   wait on gate, source'transaction, drain'transaction;
 end process;
end  switch;

entity  input_source  is
  generic(source_voltage: logic4; source_strength: strength);
  port(io_port: inout resolved_triplet);
end  input_source;

architecture  source  of  input_source  is
begin
  process
   variable state: triplet :=
                voltagestrength_to_triplet(source_voltage,source_strength);
    variable source_delay: time := ((smax-strength'pos(source_strength))+1)*d;
  begin
   io_port <= (0,0,0) after 0fs, state after source_delay;
   wait for cycle_time;
    if not io_port'active then wait on io_port'transaction; end if;
  end process;
end  source;

entity  capacitance  is
  generic(size: strength; delay_to_gate: time);
  port(io_port: inout resolved_triplet; out_to_gate: out logic4 := 'X');
end  capacitance;

architecture  capa  of  capacitance  is
begin
  process
    variable state: triplet := voltagestrength_to_triplet('X',size);
   constant source_delay: time := ((smax-strength'pos(size))+1)*d;
   constant actual_delay_to_gate: time := delay_to_gate - cycle_time;
  begin
   io_port <= (0,0,0) after 0fs, state after source_delay;
   wait for cycle_time;
    state := attenuate_triplet(io_port,size);
    out_to_gate <= triplet_to_voltage(io_port) after actual_delay_to_gate;
   if not io_port'active then wait on io_port'transaction; end if;
  end process;
end  capa;
```

Bi-directional Switches in VHDL using the 46 Value System

ALEC STANCULESCU
Fintronic USA, Inc., Willow Rd. Suite 205, Menlo Park CA 94025

Abstract. In [5] the reader can find a methodology for writing switch-level descriptions in VHDL. This paper provides an improved version of the bi-directional pass-transistor algorithm presented in [5], which is more accurate and executes fast.

Key words:
VHDL simulation, Switch-level, Distributed Bi-directional Pass-Transistor, 46 Value System.

1. Introduction

This paper continues the work presented in [3], [4], and [5], where [3] introduces the 46-value system, [4] presents the first implementation in VHDL of a bi-directional switch, as well as a methodology for implementing distributed algorithms, and [5] presents more details regarding the methodology for modeling distributed bi-directional switches, and is accompanied by a diskette containing all the relevant VHDL source code.

Two limitations of the code presented in [4], and in [5] are outlined in [1], which has been edited into this book.

This paper presents a new version of the VHDL bi-directional switch using the 46 value system presented in [5], which does not have the problems mentioned in [1], and executes faster. The interested reader should replace the nfet and pfet entities distributed on the diskette accompanying [5] with the new VHDL description presented in this paper. Because of space limitations, this paper assumes that the reader has available [5] for reference. This paper focusses only on presenting the improvements brought to the new version of the code, as well as on describing the modeling limitations.

2. Modeling Assumptions

Before presenting the algorithm, we need to understand the modeling assumptions that are made, both in terms of limitations and in terms of capabilities. Note that better models can be developed based on this methodology.

Commutation:

A switch has three pins: an input called gate, and two inout pins called source and drain.

A switch can be either on or off. A non-zero delay must be specified for the commutation time of the switch. This is necessary for the VHDL imple-

J. Mermet (ed.), VHDL for Simulation, Synthesis and Formal Proofs of Hardware, 63–71.

mentation.

Zero delay between source and drain:

The modeling assumption is that there is zero-delay between source and drain. The entire commutation time is modeled by the delay on the gate.

Superposition:

Many weak drivers may not override a stronger one.

Driving ceiling of transistors: 'R' or 'W'

A switch may not drive stronger than a certain user-specified limit. This allows the modeling of complex networks, such as the RAM cell presented in section [5].

However, only two strengths can be used: 'R', or 'W', since 'F' is reserved for power supply, and 'Z' is reserved for modeling the switching transition.

Interval-Notation Value-System:

The use of an interval-notation based value-system provides for useful simulation results by limiting unknown ('X') predictions. An example in [5] illustrates this capability.

Power supply: 'F'

No signal should be temporarily driven with the strength 'F'. Such driving should be permanent.

This limitation is due to an optimization that is based on this assumption: pass-transistors will not attempt to modify a signal with the strength 'F'. This speeds up the simulation of circuits that have many transistors connected to the same two signals: power and/or ground. Many loops are disconnected if no activity is propagated through power supply, and the simulation completes much faster.

Delays: should be multiples of 10 fs

Delays in any VHDL model that participates in the same simulation with the code of the bi-directional switch presented in this paper should be multiples of 10 fs. This will ensure that no event will affect a bi-directional switch before completion of an evaluation.

3. Improvements to the Bi-directional Switch Algorithm

The algorithm and its properties are presented in detail in [5]. The new algorithm brings the following improvements: (1) handles blocked paths, (2) does not run the risk of using an undriven dominant value during the first reconnection, and (3) improves the speed of the computation by scheduling fewer transactions.

For any particular cycle, any transistor in [5] could be : (a) untouched by the disconnection "wave", and therefore ready to disconnect, b) just disconnecting, c) active in the reconnecting process, (i.e., in the process of passing values to its neighbors), or d) inactive in the reconnecting process, (i.e., waiting for more activity or for the global activity pause). Hence, the four states of the state machine presented in [5].

The state machine representing the new algorithm consists of five states. The additional state with respect to the previous version is a state in which the transistor, having detected a potential path-blocking on one port one fs earlier, disconnects the offending port and reconnects without delta-delay, thus utilizing the neighboring dominant value that was potentially blocked.

One fs later, the neighboring transistors will accept and propagate the blocked value since one fs later they will be ready to disconnect and thus ready to discard any undriven value latched on the path.

The condition that warns of a potential blocked paths is that the new value passing through signals DRN or SRC (a) has a state different from 'X', as well as different from the current value, and (b) is stronger than the current value, which in turn is strictly stronger than "Z". The code saves the information whether the blocking occurs on DRN or SRC.

The state in which the disconnecting is performed is enhanced to wait for two delta cycles before reconnecting. The first wait is used to not be woken up by its own disconnection, and the second wait is used to not be woken up by to the disconnection of the neighboring transitors, which will supply useless disconnection information.

In order to detect whether to disconnect or to reconnect, the simulated time is saved during disconnection and only one disconnection is permitted per timepoint. Therefore, all subsequent process invocations will result in reconnections.

Note that in order to handle blocked paths the computation may require more than one femtosecond. Given that there are two strengths in the 46-value system that can be used to limit the driving capability of transistors (i.e. 'R', and 'W') a maximum of two femtoseconds are required to complete the algorithm.

The VHDL code corresponding to the core algorithm is presented in the next section.

4. Description of the VHDL Implementation

4.1. VHDL Code

The new version of the VHDL bi-directional pass-transistor is presented below. It depends heavily on the functions of the std_logic package, which is presented in [2], and in [5].

```
USE work.STD_LOGIC.ALL;
-------------------------------------------------------
-- nfet: Bidirectional XFR gate with non-inverting gate
-------------------------------------------------------
--
entity nfet is
    generic (gdelay : time := 3 ps; MaxStrength : t_strength := 'R')
        --  gdelay       : delay on the gate G
        --  MaxStrength : max output strength on Src and Drn.
        --                Used to model resistance. Its value
        --                should be either 'R' or 'W'
    port( g: in t_wlogic; src,drn: inout t_wlogic);
        -- Src and Drn are symmetric.
end nfet;
--
architecture nfet_behavior of nfet is
        signal gd : t_wlogic;                -- delayed (internal) gate
        signal again : boolean := false; -- for blocked paths
begin
    gd <= g after gdelay;
    process
        variable psrc, pdrn : t_logic;    -- to store old contribitio
        variable vsrc, vdrn : t_logic;    -- to store next contributi
        variable init : boolean := true; -- to initialize at powerup
        variable first : boolean := true; -- to disconnect at poweru
        variable last_time : time:= 0;    -- to check for quiescence
        variable blocked_src, blocked_drn: boolean := false;
  begin
    if again and (not (NOW = 0 fs)) then
        again <= false;
        if blocked_src then
          blocked_src := false;
          if f_strength(src) /= 'F' then
            src <= vsrc;
            last_time := NOW;
          end if;
        end if;
        if blocked_drn then
          blocked_drn := false;
          if f_strength(drn) /= 'F' then
            drn <= vdrn;
            last_time := NOW;
          end if;
        end if;
    else
       if init then
         vsrc := D;
         vdrn := D;
         psrc := D;
```

```
      pdrn := D;
      src <= D;
      drn <= D;
      init := false;
      last_time := 0 fs;
    elsif (last_time < NOW) or ((NOW = 0 fs) and first) then
      last_time := NOW;
      first := false;
      if (gd'event or (f_state(gd) /= '0')) then
        vsrc := f_convz(src);
        vdrn := f_convz(drn);
        psrc := vsrc;
        pdrn := vdrn;
        if f_strength(src) /= 'F' then
          src <= vsrc;
        end if;
        if f_strength(drn) /= 'F' then
          drn <= vdrn;
        end if;
        wait for 0 fs; wait for 0 fs;
      end if;
    end if;

    case f_state(gd) is
      when '0' =>
        if src'active or gd'active then  -- kick the src back
          vsrc := f_convz(src);
        end if;
        if drn'active or gd'active then  -- kick the drn back
          vdrn := f_convz(drn);
        end if;
      when '1' =>
        if f_strength(drn) < f_strength(src) then
          vdrn := f_ceil(src,MaxStrength);
          vsrc := f_convz(vdrn);
        elsif f_strength(drn) > f_strength(src) then
          vsrc := f_ceil(drn,MaxStrength);
          vdrn := f_convz(vsrc);
        else
         -- f_busres computes the resolved value
          vdrn := f_ceil(f_busres(src)(drn),MaxStrength);
          vsrc := vdrn;
        end if;
      when 'X' =>
         -- f_uxfr computes the range between tristate
         --         and the resolved value
        if f_strength(drn) < f_strength(src) then
          vdrn := f_ceil(f_uxfr(FX,src,f_convz(drn)),MaxStrength);
          vsrc := f_convz(vdrn);
        elsif f_strength(drn) > f_strength(src) then
          vsrc := f_ceil(f_uxfr(FX,drn,f_convz(src)),MaxStrength);
```

```
          vdrn := f_convz(vsrc);
        else
          vdrn := f_ceil(f_uxfr(FX,f_busres(src)(drn),
                                f_convz(drn)),MaxStrength);
          vsrc := f_ceil(f_uxfr(FX,f_busres(src)(drn),
                                f_convz(src)),MaxStrength);
        end if;
      end case;

      if psrc /= vsrc then  -- update when change
        if ((f_strength(vsrc) >= f_strength(psrc)) and
             (f_strength(psrc) > 'Z') and
             (f_state(vsrc) /= 'X') and
             (f_state(vsrc) /= f_state(psrc))) then
          blocked_src := true;
          again <= true after 1 fs;
        end if;
        psrc := vsrc;
        if f_strength(src) /= 'F' then
          src <= vsrc;
        end if;
      end if;

      if pdrn /= vdrn then  -- update when change
        if ((f_strength(vdrn) >= f_strength(pdrn)) and
             (f_strength(pdrn) > 'Z') and
             (f_state(vdrn) /= 'X') and
             (f_state(vdrn) /= f_state(pdrn))) then
          blocked_drn := true;
          again <= true after 1 fs;
        end if;
        pdrn := vdrn;
        if f_strength(drn) /= 'F' then
          drn <= vdrn;
        end if;
      end if;
    end if;
    wait on gd,src'transaction,drn'transaction, again;
    end process;
end nfet_behavior;
```

Note that the description of pfet (pass-transistor with inverting gate) is similar to the description of nfet. To obtain the pfet description from the nfet description the following two steps are required: (1) replace all occurrences of nfet with pfet, and (2) replace the assignment to gd (there is only one) with the assignment presented below:

```
gd <= f_logic(f_not(f_state(g)))('R') after gdelay;
```

.2. Comments about the Entity Declaration

The bi-directional switch is described as an entity with two generic parameters (Gdelay, and Maxstrength), and three external ports G, SRC, and DRN corresponding to gate, source, and drain. Gdelay represents the delay on the gate and models the speed of commutation. MaxStrength represents the maximum strength with which the switch may drive a signal connected at its drain or source and models the voltage drop between source and drain. Its use is exemplified in [5]. Note that the values of MaxStrength must be either 'R' or 'W'.

.3. Comments about the Architecture Declaration

The architecture represents the execution code of the model. It consists of two concurrent statements: a delayed assignment and a process.

The delayed assignment of G, to the internal signal GD ensures that a transaction on GD can occur only during the first simulation cycle of a timepoint.

4.4. Comments about Variables in the Main Process

Variables are used to store the values previously driven by the process (PSRC, and PDRN), as well as the values that have to be driven next (VSRC, and VDRN). The same values are not assigned again in during reconnection. This ensures the completion of the process activations for a particular time point.

Variable LAST_TIME is used to store time and to ensure that a transistor disconnects on once during each timepoint.

Variable init, is used to initialize signals and varaibles. Variable First is used to compensate for the first disconnection, when variable LAST_TIME has the value 0, which is the same as the value returned by the function NOW at that time.

4.5. Comments about the Executable Part of the Main Process

The process is sensitive to events on signals GD, SRC'transaction, DRN'transaction[1], and AGAIN. When GD changes its value, the values on SRC and DRN need

[1] Note that S'transaction is a signal that has events whenever S has transactions. Any change in the value of a driver of a signal S represents a transaction on S. A transaction on S may not result in an event on S, because the resolved value of all drivers of S may produce a value equal to that of S.

to be recomputed in response to the commutation of the switch.

It is important to use S'transaction because processes in VHDL are sen sitive only to events on signals, and the model of the bi-directional switc must react to transactions rather than events on SRC and DRN when prop agating the disconnection along a path of open switches.

The process is sensitive to signal AGAIN in order to unblock paths tha are potentially blocked.

In the disconnecting state, the pass-transistors remove their driving e fects. In the reconnecting state the pass transistors "pass", in both direc tions, the resolved values of the driving sources. Note that pass transistor do not "drive" signals, they just "pass" the values from the driving source The driving sources come from power, ground, or external gates.

4.6. Optimizations

Some optimizations were used in the new version of the code:

1) In order to decrease the number of process evaluations, the new algo rithm takes advantage of the direction of propagation. For example if SRC is stronger than DRN, the transistor will drive DRN with the resolved value and SRC with the same state as DRN but with strength 'Z'. In other words if DRN = W0, and SRC = R0, then the transistor will drive DRN with R and SRC with Z0.

2) All assignments to signals, except for initialization (since the valu U, of uninitialized signals is of power supply strength), check whether th strength is that of power supply. This optimization helps in large network connected to the same vdd and gnd signals.

3) Common subexpression elimination has been performed on assigne ments to VSRC and VDRN.

5. Conclusion

This paper presents an improved version of the bi-directional transistor pre sented in [5]. In conjunction with the 46-value system Standard_Logic pack age, placed in public domain by Vantage Analysis Systems, the code shown in this paper can be used to integrate in the same simulation both high-leve and switch-level VHDL descriptions. Furthermore, the VHDL descriptions are portable, and the resulting simulations are of acceptable speed.

6. References

[1] K. Khordoc, M. Biotteau, E. Cerny, *Switch-Level Models in Multi-Leve VHDL Simulations*, Proceedings of the First European Conference on VHDL, Marseilles, Sept. 1990.

[2] D.R. Coelho, *The VHDL Handbook*, Kluwer Academic Publishers, 1989.

[3] A.G. Stanculescu, A.S. Tsay *A 46-State Value-System from Vantage Analysis Systems, Inc.*, VDEG, February 89.

[4] A.G. Stanculescu, A.S. Tsay, A.N.D.Zamfirescu, D.L. Perry *Switch-Level VHDL Descriptions*, ICCAD89.

[5] R. Harr, A.G. Stanculescu *Applications of VHDL to Circuit Design*, Kluwer Academic Publishers, 1990.

Acknowledgements

Andy Tsay, Alex Zamfirescu, and Doug Perry contributed to [3] and [4], which represent the basis for this [5]. Many thanks go to Vantage Analysis Systems for placing the 46-value system in the public domain, along with the entire accompanying package (std_logic).

Systems Real Time Analysis with VHDL Generated from Graphical SA-VHDL

MATTI SIPOLA*, JUHA-PEKKA SOININEN**, JORMA KIVELÄ*

*) *University of Oulu, Electronics Laboratory, SF-90570 Oulu, Finland*
**) *Technical Research Center of Finland, Electronics Laboratory, Oulu*

1 Abstract

This paper will demonstrate a method in which the high level system specification is defined using an user-friendly CASE tool together with VHSIC Hardware Description Language, VHDL. This method called SA-VHDL is especially suitable for system partitioning process of digital real-time embedded systems. A prototype tool SYS-RTA has been implemented to demonstrate the automatic conversion from the CASE tool output to the executable analysis model in VHDL. Simulation of the generated VHDL model reports the performance, i.e. response times, the resource usage, function activities and behavioural errors of the system. These results can be used for the quality analysis of the current system partitioning.

2 Introduction

Due to recent semiconductor technological advances it is possible to design a complex embedded system with processor, software and hardware logic on one chip. In order to avoid integration problems the

J. Mermet (ed.), VHDL for Simulation, Synthesis and Formal Proofs of Hardware, 73–86.

behaviour of the system must be very carefully verified and analysed at the specification level. Special methods and tools are required to the development of an executable specification that can be thoroughly evaluated before implementation is begun. Timing requirements from the outside world has a strong influence to the system partitioning. System designer must be able to rapidly evaluate and compare affects of different software/hardware resource[1] selections. Unexpected queuing of resource leads usually to violation of response time constraints. With optimum performance and use of resources the real-time operation is guaranteed and the cost of the system can be kept as low as possible.

Some commercial CASE tools can evaluate real-time system behaviour by means of animation or high level simulation. CARDtools [1] supports timing analysis of a software specification on a selected microprocessor with well defined hardware/software interfaces. Critical paths can be evaluated for a selected operation system by estimating the time required for yet uncoded tasks. Teamwork/SIM [2] has the ability to simulate real-time systems at higher and more implementation independent level. It is suitable for complex system-architecture modelling issues, such as partitioning, allocation, process-to-resource mapping and performance estimates. Performance can be evaluated with estimated delays of processes and stores and parameters of microprocessors. The main drawback of these tools is that they are software oriented and concentrate only to simulate software on some microprocessor hardware. Simulation language is not standardized and specifications are not suitable for hardware design.

Another approach to verify embedded systems is to simulate software running on a real CPU hardware at behavioural or register level. This kind of simulation is strongly implementation dependent and possible only after software and hardware parts have been designed. Timing constraints for a real-time system are not practical to test because of the long simulation runs. Hardware simulation method is more suitable for accurate timing verification of software/hardware interfaces as the last design phase prior to the first prototypes.

Standardized hardware description language VHDL [3] has been adopted also to the system specification work. SpecChart is a special

---

1. With the resource we do not mean only the processor, but also other costly hardware devices e.g. ALU or multiplier, which can handle only one function at the same time, but can be shared to set of functions by scheduling.

graphical method for generating executable VHDL specifications [4]. It is capable of specifying complex controlling schemes of real-time systems. The UVA uninterpreted modelling system works at a higher level of abstraction [5]. Method is used for evaluating system performance and allows function of blocks to remain undefined. This helps to find delay constraints for undefined functions to satisfy timing dead lines. Graphical method used on both of these systems seems not to be suitable for system partitioning process.

Totally implementation independent approach is required for an optimum partitioning of a real-time embedded system. Functions must be allocatable to different resources or to parallel hardware logic as long as possible. Method must be suitable for specifying complex, hierarchical and parallel real-time systems. A new system specification method, called SA-VHDL, is designed especially for digital real-time embedded systems. SA-VHDL is a graphical CASE tool based method for system analysis work with VHDL. It is based on Ward & Mellor's Real Time Structured Analysis (RT/SA) method [6] and Ward's Transformation Schema [7]. Typical set of SA-VHDL diagrams is shown in figure 1. VHDL have been adopted as a textual part of the method. VHDL programming has been made as easy as possible. User has to enter only the behavioural parts of the processes and VHDL data definitions to data flows. Both hardware and software designers are capable for this work. The design platform for the SA-VHDL method can be constructed of both commercial or academic software. Non-commercial prototype tool for a special SA-VHDL graphical entry has been already implemented [8].

SA-VHDL concentrates to specify functions on system's behavioural level. These functions can be allocated to different implementations. After the software/hardware selections have been done an executable VHDL model with resource controllers can be generated. Overall performance is highly dependent of possible queuing of resources. Total execution times of the functions can unexpectedly exceed the timing constraints. Computer assistance is required for an iterative method of finding the best possible partitioning of the system. Earlier experiences with automated use of SA and VHDL have been presented, but any resource modelling and timing aspects have not yet been concerned [9][10].

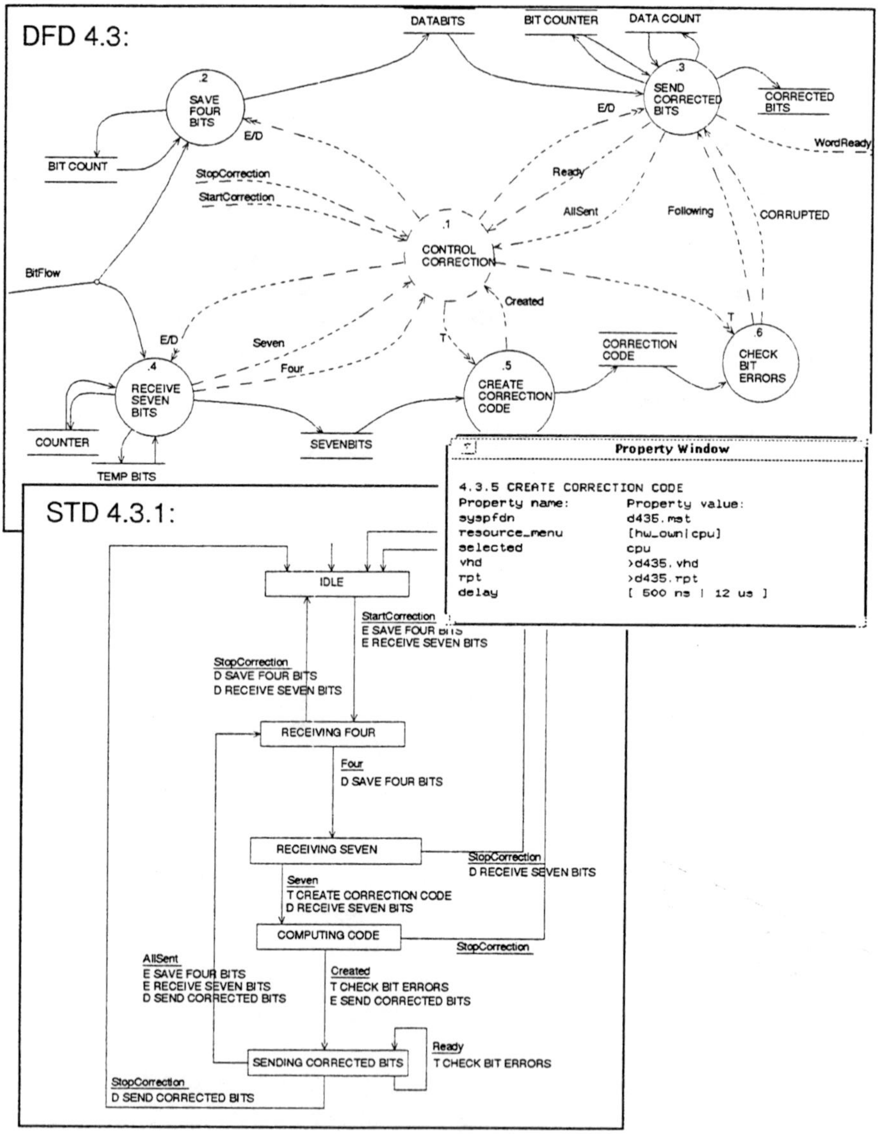

Fig. 1. Typical pair of lowest level SA-VHDL diagrams. Parameters for one of the DTRs are shown.

A prototype generator SYS-RTA has been implemented to demonstrate the automatic conversion from the graphical SA-VHDL resource model to the executable analysis model in VHDL. In addition to normal verification of behaviour, VHDL analysis model reports following items: performance, function activity, resource usage and errors on data flow. Code generation to an abstract VHDL model gives a fast iteration cycle to find out a correct software/hardware partitioning of the system with an optimum performance/cost ratio. Experiences with a case study are discussed at the end of this paper.

3 Designing system's resource model with SA-VHDL

3.1 Overview of the SA-VHDL method

SA-VHDL specification consists of hierarchical data flow diagrams (DFDs) controlled with state transition diagrams (STDs). Typical pair of a data flow and its controlling state transition diagram on lowest hierarchical level is shown on figure 1. Data transformations (DTRs, solid bubbles in figure 1) are defined with an one-active-input VHDL processes or with a new DFD. Elements with dashed lines in the figure 1 represent the control of the system.

With SA-VHDL method the system can be specified with four phases. First the *essential model* of the system must be specified. It is totally implementation independent and is concerned only on essential behaviour of the system. Interfaces to outside world are abstract and show only the logical information flow between system and system's users. Essential model is extended to *functional model* by specifying system's interfaces as they are or will be implemented. This means that information flow is more accurately defined and some of the DTRs must be refined with new DFDs and some new DTRs must be added to the system's interfaces. DTRs are defined to have performance requirements to satisfy system's port to port timing deadlines. *Resource model* is derived from functional model by allocating DTRs to different implementation resources. Output delays to DTRs must be specified at this phase. Resource model syntax doesn't allow any use of data in the state transition diagrams. Possible use of variables must be defined with new DTRs. Understanding of the system is easier, because specifications of data and control are efficiently separated on every DFD. It is very visible how separate real-time events affect on individual data transformations. Final *implementation models* can be constructed by dividing the SA-VHDL architecture to separate SA-VHDL specifications of different resources and parallel hardware.

3.2 Tools for the SA-VHDL method

We have adopted a commercial CASE tool Prosa [11] to our interface for SA-VHDL specifications. The template of the VHDL minispecification for a DTR can be generated with Prosa by pushing to the text editor from the DTR bubble in the DFD. Input/output flow matches to lower level diagram or to VHDL code are checked automatically on every hierarchical push from upper level. All data elements (e.g. flows and stores) can be defined with predefined VHDL data types or with data types from an user defined package. Selections of the resources and output delays to data transformations are given with parameters. Our compiler SYS-RTA can be started directly from Prosa and it uses Prosa's prolog net list output as an input database.

3.3 The SA-VHDL resource model

SA-VHDL resource model *syntax* is based on RT/SA. Some new elements e.g. merges and flags have been added to diagrams. Each lowest level data transformation can be allocated to different resource, e.g. CPU, multiplier, ALU or to parallel hardware implementation.

Execution rules of the SA-VHDL resource model are based on Ward's Transformation Schema. Use of resources has added some features. Each resource can be used by only one DTR process at the same time. This means that the resource must have a scheduler and every DTR allocated to it must have an opposite handshake controller. Control signals between functions and resource have no delay in the resource model. All the delays in the model exist only for the DTR output flows, inside the execution model of the DTR.

3.4 Simulating the resource model with VHDL processes

For each DTR bubble we have a state diagram shown in figure 2.[1] Following the states, the DTR process can be executed only when the DTR gets to "Active" state. If the DTR has to wait for the resource, this behaviour cannot be simulated without some additional processes. These

1. This behavior is not defined by the user. User has to define only the DTR behavior inside the generated VHDL process templates.

processes are the latch process for catching the activating input event and the execution controller process derived from the state diagram in the figure 2. Separate delay processes for outputs can handle a situation, where DTR has been disabled during its execution. Outputs are not produced and receiving DTRs do not get activated. The resource scheduler process can be implemented with many different scheduling algorithms. Simplest of them is a polling method, where service request signal for each DTR is checked in sequence and resource is reserved for the requesting DTR as long as it is executed or disabled.

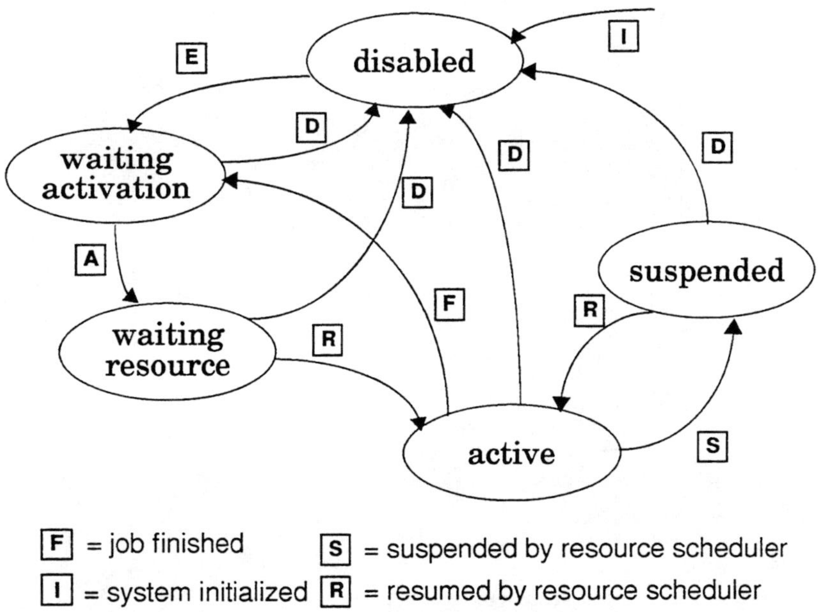

Fig. 2. Functional States for a Data Transformation Bubble (DTR).

Resource model can be analysed with some of the DTRs undefined. *Uninterpreted modelling* is most suitable for evaluating system's overall performance on very early phase of specification work. If the data values on outputs of the DTR could not affect the timing behaviour of the system, the behaviour of the DTR can be kept uninterpreted. DTR activity is simulated by assigning a valid value to each output with estimated output delays. For each VHDL data type, a value to represent the absence of information must be defined. All other values are interpreted as tokens i.e. valid packets of information. Figure 3 shows an example of a generated uninterpreted data transformation specification for DTR 3 in figure 1. If the correct data transformation function is needed, the process "Uninterpreted" can be replaced with the real behaviour specified with same signal names.

```
-- Data Tr: 4.3.3 SEND CORRECTED BITS
-- ==================================
--
USE Work.Defs.all;
--
ENTITY d433 IS
   PORT
   (
   DATABITS:IN             DATABITS_sys_rta_type;
   Following:IN            Following_sys_rta_type;
   CORRUPTED:IN            CORRUPTED_sys_rta_type;
   ExD:IN                  ExD_sys_rta_type;
   DATA_COUNT:IN           DATA_COUNT_sys_rta_type;
   BIT_COUNTER:IN          BIT_COUNTER_sys_rta_type;
   WordReady:OUT           WordReady_sys_rta_type;
   CORRECTED_BITS:OUT      CORRECTED_BITS_sys_rta_type;
   Ready:OUT               Ready_sys_rta_type;
   AllSent:OUT             AllSent_sys_rta_type;
   DATA_COUNT_out:OUT      DATA_COUNT_sys_rta_type;
   BIT_COUNTER_out:OUT BIT_COUNTER_sys_rta_type

   );
END d433;
-- ==================================================
ARCHITECTURE Behaviour OF d433 IS
BEGIN
   UnInterpreted :
   PROCESS (Following'TRANSACTION)
   BEGIN
      WordReady <= S ;
      CORRECTED_BITS <= B"1111111111111111" ;
      Ready <= S ;
      AllSent <= S ;
      DATA_COUNT_out <= 1 ;
      BIT_COUNTER_out <= 1 ;
   END PROCESS;
END Behaviour;
```

Fig. 3. Generated DTR specification template.

4 Analysis of the resource model with VHDL

A translator from SA-VHDL to VHDL analysis model has been implemented. This prototype tool SYS-RTA (SYStems Real Time Analyser) has been programmed in IF/Prolog and tested on SUN4 and UNIX. Figure 4 shows the main features of generated VHDL. In the analysis model each primitive level graphical SA-VHDL object has an error and performance reporting VHDL process. Scheduler process for each shared resource and execution controller for each data transformation function are also generated. By simulating the model with hardest real-time stimulus valuable analysis data can be read directly from the generated report signals. In addition to normal verification of behaviour, VHDL analysis model reports following items: performance, function activity, resource usage and errors on data flow. All these reports have been selected, since the information is relevant in the partitioning process and verification of the model. The analysis methods are especially designed for estimating software/hardware partitioning quality of embedded systems [12].

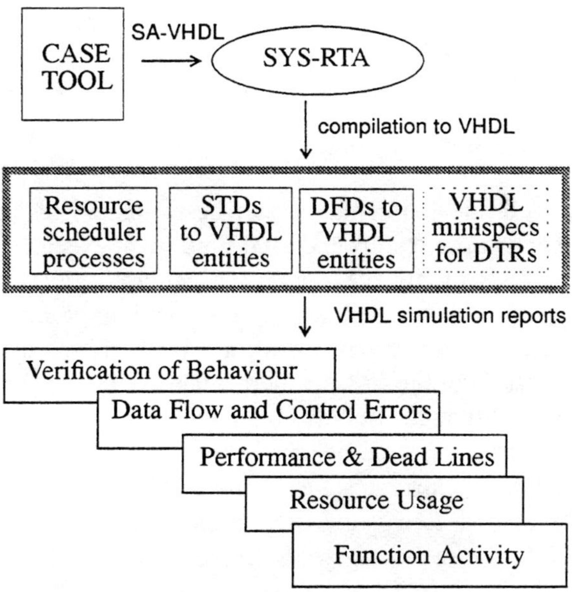

Fig. 4. Analysis functions of generated VHDL.

The *performance report* must be generated for each response time requirement. Report is made for each input event if it has or has not generated expected output event in a given port to port time. Individual DTRs can have also overall delay requirements which must be checked. Both of these reports have been implemented in the analysis model with separate VHDL processes. Example of a generated VHDL process for a response requirement is shown in figure 5. VHDL has ability to read current simulation time with function NOW. Required timing calculations can be performed during simulation. With the transaction attribute of a VHDL signal an event with same data value as old one can be noticed.

```
-- ==========================================
-- Port to port time recorder :
--
BitFlow_2_WordReady : PROCESS
   VARIABLE pp_t, pp_in : Time;
BEGIN
   WAIT ON d_bitflow'TRANSACTION ;
   pp_in := NOW ;
   WAIT ON s_wordready'TRANSACTION for 16*30 us * 2;
   pp_t := NOW - pp_in;
   BitFlow_to_WordReady <= pp_t;
   IF pp_t < 16*30 us THEN
      ASSERT FALSE
      REPORT "RESPONSE_OK AT s_wordready"
      SEVERITY Note ;
   ELSE
      ASSERT FALSE
      REPORT "RESPONSE_NOT_OK AT s_wordready"
      SEVERITY Warning ;
   END IF;
END PROCESS;
```

Fig. 5. Process for a response requirement.

For each DTR the performance can be estimated by comparing the time between activating events and total time spent executing the DTR. In the VHDL analysis model these *function activity reports* are generated inside the execution controller, which is a common component for each DTR. The report is implemented with a VHDL record signal from the data transformation bubble. It has a percentage value for showing ratio of being active and activated. Report also includes time signals telling how much function has spent time being disabled, waiting activation, waiting resource and active. Current state of the DTR function is shown with previous state names as a value of a enumerate signal. State report of DTR functions in figure 1 is shown in figure 6.

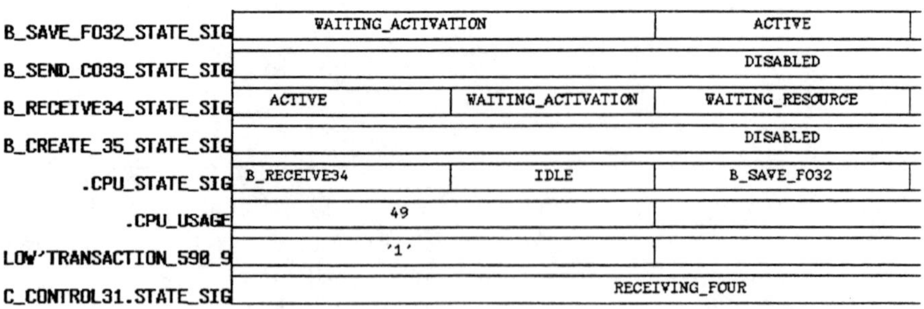

Fig. 6. DTR function and resource state reports for the DFD in figure 1.

The primary job of the *resource usage report* is the verification of resource capacity with heaviest input loading of the system. In the analysis model it is generated with VHDL signals and assert clauses. SYS-RTA generates resource scheduler for each common resource (e.g. CPU, multiplier, ALU). Resource scheduler communicates by using hidden VHDL signals with execution controllers of data transformation bubbles. Report signals shows the name of the function which is using resource and the percentage ratio of resource being active compared to simulation time spent. Competitions on queue of resource are reported with VHDL assert clauses. Typical resource usage report is shown in figure 7.

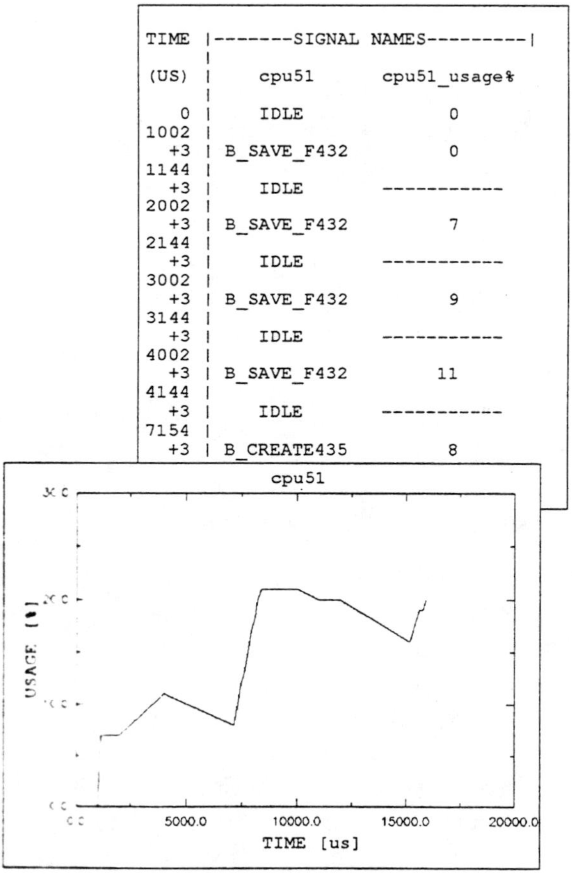

```
TIME |-------SIGNAL NAMES---------|
     |
(US) |     cpu51      cpu51_usage%
     |
   0 |     IDLE                 0
1002 |
  +3 | B_SAVE_F432              0
1144 |
  +3 |     IDLE       -----------
2002 |
  +3 | B_SAVE_F432              7
2144 |
  +3 |     IDLE       -----------
3002 |
  +3 | B_SAVE_F432              9
3144 |
  +3 |     IDLE       -----------
4002 |
  +3 | B_SAVE_F432             11
4144 |
  +3 |     IDLE       -----------
7154 |
  +3 | B_CREATE435              8
```


Fig. 7. Resource usage reports.

Generated analysis model contains many different *data flow and control error reports*. From data flow diagrams they report e.g. data crashes on stores, losing of data or control on transformation inputs and resource competition situations. From state transition diagrams competitive true conditions for a state transition are reported. If an oscillation inside control transformation is detected simulation is terminated and the place where oscillation occurred will be reported. Example of a generated analysis model of a store is shown in figure 8. *Last_active*-attribute of the VHDL signal tells the amount of time that has elapsed since the last time the store was updated. For too narrow updates of the store an error report is produced.

```
-- ==========================================
-- SA Element: 3 COUNTER sysosto
--
STORE_t_counter3 : PROCESS
BEGIN
    WAIT ON i_counter'TRANSACTION ;
    IF i2counter'Last_Active > 0 ns AND
        i2counter'Last_Active
        < SYS_RTA_DEFAULT_RESOLUTION_TIME
    THEN
        ASSERT FALSE
        REPORT
        "DATA_CRASH at Store i2counter " &
        "- resolution time violated "
        SEVERITY Warning;
    END IF;
    i2counter <= i_counter;-- input to store output
END PROCESS;
-- ==========================================
```

Fig. 8. Generated error reporting process for a store.

5 Experiments with SYS-RTA generator

SYS-RTA generator has been tested with several different embedded systems. Largest of them is an embedded digital signalling system in a communication system. The SA-VHDL specification consists of 14 data flow diagrams, 12 state transition diagrams and 39 data transformation specifications with 2800 lines of VHDL code. This example and the manual coding of its functional specification is presented in [13]. SYS-RTA translated the SA-VHDL specification to 12,500 lines of VHDL with all analysis functions in 6 minutes. Manual coding from the same specification to an executable analysis model has been estimated to take about two weeks. Simple parts of the system with one or two diagrams have been translated to VHDL in few seconds. With the same compiler technology one STD can be compiled to a VHDL input for a hardware synthesis system in few tenths of second.

Simulation of example's resource model helped to find out bottle necks, which were not visible from the hierarchical diagrams. Allocating DTRs to same resource in different DFDs led to unexpected queuing, which resulted to a response time violation. These weak points were found by reading state reports of DTR execution controllers. Signal value "Waiting Resource" was easily found from the textual simulation results.

Generated VHDL analysis model was simulated with Intermetrics [14] simulator. Same models were simulated also with PC based V-System/ Windows [15]. Both of these simulators produced textual simulation results, which were easy to convert to graphical format with simple filter programs.

6 Conclusions

VHDL have proved to be very efficient language for system level analysis functions. Because VHDL is quite cumbersome to write, lots of computer assistance in coding large VHDL analysis models is required. Current commercial CASE tools are very helpful for embedded software coding. They can be adopted also as a new interface for digital system specification work with VHDL. Experiences with SA-VHDL and SYS-RTA generator have proved that the system's behavioural specification can be generated to an executable resource model in VHDL. Generation of VHDL can be highly automatized. Only 10-20% of the final analysis code must be manually entered inside automatically generated VHDL templates. These behavioural parts are easy to code by both software and hardware designers. Analysis reports of the system's behaviour and performance produce a lot of valuable information to system partitioning process.

REFERENCES:

[1] Schindler, M., Design tools master embedded software's real-time demands, Electronic Design, Sept. 17 1987, pp. 65-66

[2] Maliniak, L., One tool verifies system and software, Electronic Design, Nov. 22 1990. pp. 49-52

[3] IEEE Standard VHDL Language Reference Manual, IEEE Std 1076-1987, N.Y. 1988

[4] Narayan, S. & Vahid, F. & Gajski, D.: Translating Systems Specifications to VHDL, The European Design Automation Conference, EDAC 91.

[5] Hady, F.T., Aylor,J.H., Waxman, R.: Uninterpreted modeling using the VHSIC hardware description language (VHDL), ICCAD 89, Santa Clara, CA.

[6] Ward, P.T. and Mellor, S.J., Structured Development for Real-Time Systems, Vol. 1-3. Yourdon Press, N.Y. 1986

[7] Ward, Paul T., The Transformation Schema: An Extension of the Data Flow Diagram to Represent Control and Timing, IEEE Transactions on Software Engineering, Vol SE-12, No. 2, Feb86

[8] Lahti, J., Sipola, M., Kivelä, J., SADE: A Graphical Tool for VHDL-based System Analysis, ICCAD 91, Santa Clara, CA., U.S.A., Nov. 11-14 1991

[9] Tikkanen, T., Leppänen, T., Kivelä, J., Structured Analysis and VHDL in Embedded ASIC Design and Verification, EDAC 1990, Glasgow, pp. 107-111

[10] Kivelä, J, Leppänen, T., Sipola, M., VHDL Supported by Graphical CASE Tools for High Level System Design & Implementations EUROVHDL90, Marseille, France 1990

[11] Tervonen M., Prosa - Professional Structured Analysis Tool for Software Development, Daisee87 symposium, Oslo, Norway, Nov87

[12] Juha-Pekka Soininen, Matti Sipola, Kari Tiensyrjä: SW/HW-Partitioning of Real-Time Embedded Systems, Microprocessing and Microprogramming, The Euromicro Journal, Volume 27, No. 1-5, 1989, pp. 239-244.

[13] Matti Kauppi, Juha-Pekka Soininen: Functional Specification and Verification of Digital Systems by Using VHDL Combined with Graphical Structured Analysis (SA), EUROVHDL 91

[14] VHDL Design Environment (VDE) User's Manual, Version 3.0 (Unix), 21 December 1990, Intermetrics

[15] V-System/Windows User's Manual - VHDL for Microsoft Windows 3.0, Model Technology, 1991

Delay Calculation and Back Annotation in VHDL Addressing the Requirements of ASIC Design

Peter Connor
Sanjay Nayak
Joyce Kraley
Victor Berman

Cadence Design Systems, Inc.
CAE Division
2 Lowell Research Center Drive
Lowell, MA 01852 USA
508 458 1900
berman@cadence.com

Abstract

The timing model in VHDL was developed to allow very sophisticated modelling of complex inter-process interactions that occur at the system level. The unique characteristics of this model allow timing computations to be freely mixed between behavioral and structural elements of the design. However, this flexible model creates problems for typical ASIC designers since there is no industry accepted standard practise for the representation of timing data in VHDL models or libraries. Furthermore, language restrictions make delay calculations that are dependent on connectivity data cumbersome and inefficient while concepts that are familiar to ASIC designers, such as pin-to-pin timing, are difficult to express in VHDL and generally alien to the underlying timing model of the language.

Cadence has been working on a methodology to solve this problem in a straight forward way. It's a methodology that allows accurate timing data to be shared by all VHDL simulators and allows ASIC designers to continue using their proven methods of delay calculation. This methodology has three basic components:

- standard multi-value logic (MVL) package
- standard format for storage of delay data
- standard "access routines" for VHDL

A standard MVL is needed since delays are generally dependent on logic level transitions. Without a common nomenclature and semantic for these levels it would be impossible to clearly denote these transitions. Significant progress in this area makes these ideas a practical reality.

J. Mermet (ed.), VHDL for Simulation, Synthesis and Formal Proofs of Hardware, 87–98.

A standard delay format is needed so that all simulators may have access to the same data. A neutral format has been developed and will be presented in this paper. There are plans to submit this format to both IEEE and CFI as a strawman standard.

The proposed VHDL access routines implement the methodology.

The Problem

Delay calculation and back annotation are typical applications where data, in this case delay values, is specified in an external file that then needs to be incorporated into the user's design. Generally, delay values are included as part of a component's generics list. This approach allows these values to be set on an instance by instance basis. A mechanism does not currently exist in the language to support initialization of generics from reading a file.

A second problem is the need to use hierarchical names in the delay file to identify the argument to be updated. Currently, hierarchical name referencing is not supported in VHDL.

A third problem is the need to document the values in this external file in the source code. Today, this is basically a manual effort which is prone to human error. For ease of updating source code and correctness, an automated mechanism is desirable.

Cadence has been working on a methodology to solve this problem in a straight forward way given the existing language limitations. This methodology has three basic components:

- standard multi-value logic (MVL) package
- standard format for storage of delay data
- standard "access routines" for VHDL

With some enhanced language semantics and a couple of new constructs, timing information can be specified in a more concise and efficient manner. A side benefit of these proposed changes is that they clarify the functionality to be performed, as well as separating the technology-specific data from the functionality.

Standard Multi-Value Logic Package

For ease of use and consistency, the standard MVL package would minimally include the type definition for the MVL plus subtypes, resolution functions, and appropriate overloaded operators. The proposed IEEE multi-value logic type definition is as follows:

```
TYPE std_ulogic is
      ( 'U', -- Uninitialized
      'X', -- Forcing 0 or 1
      '0', -- Forcing 0
      '1', -- Forcing 1
      'Z', -- High Impedance
      'W', -- Weak 0 or 1
      'L', -- Weak 0 ( for ECL open emitter )
      'H', -- Weak 1 ( for open Drain or Collector )
      '-' -- don't care
      );
```

By having a standard logic type, the nomenclature and semantics of logic level transitions can be clearly denoted. Consequently, the correct propagation delay values can be applied based on the transition levels thus allowing for a more accurate simulation model. An overloaded function could be provided in the package to identify the delay value to use based on the `'last_value` and the current value for clarity of the intended chip functionality.

Standard Format for Storage of Delay Data

A standard delay file simplifies the number of utilities a vender needs to update the appropriate information in the design—whether for simulation or synthesis. For the user, this results in a common file that is portable.

Cadence has proposed a Standard Delay File (SDF) format that can be used by a number of tools and applications. Figure 1 illustrates the kind of information that would be put in the SDF.

(DELAYFILE
<design name>
<date>
<vendor>
<program name>
<version>
<hierarchy_divider>
<voltage>
<process>
<temperature>
<time scale>
<cell>?
)

where <cell>? means that multiple cell sections are allowed.

Figure 1: Standard Delay File Format

An example of an SDF file for a 4-bit register design that contains 4 D flip-flops plus an AND gate is provided below. To see the connectivity more clearly, refer to the graphical depiction of the circuit shown in Figure 2.

```
(DELAYFILE
  (DESIGN "4-bit Register")
  (DATE "Mon July 22 10:27:33 EST 1991")
  (VENDOR "Cadence")
  (PROGRAM "back_annotate")
  (VERSION "1.1")
  (DIVIDER /)
  (VOLTAGE 4.5:5.0:5.5)
  (PROCESS typical)
  (TEMPERATURE 55:85:125)
  (TIMESCALE 1 ns)
```

```
(CELL
  (CELLTYPE "DFF")
  (INSTANCE *)
  (DELAY
    (ABSOLUTE
      (IOPATH d q (4 : 5.5 : 7)
             (4.5 : 6 : 7.5))
      (IOPATH ena q (3 :: 3) )
      (IOPATH rst q (2.5 :: 2.5)))))
(CELL
  (CELLTYPE " ")
  (INSTANCE "R1/A1")
  (DELAY
    (ABSOLUTE
      (IOPATH in1 y (2 :: 2.3))
      (IOPATH in2 y (2 :: 2.3)))))
(CELL
  (CELLTYPE "REGISTER4")
  (INSTANCE "R1")
  (DELAY
    (ABSOLUTE
      (INTERCONNECT R1/D1/q R1/D2/d
        (.145 :: .145)(.125 :: .125))
      (INTERCONNECT R1/A1/y R1/D1/clk
        ( .135:: .135) (.125 :: .125))
      (INTERCONNECT R1/A1/y R1/D2/clk
        (.140 :: .140) (.130 :: .130))
      (INTERCONNECT R1/A1/y R1/D3/clk
        ( .145:: .145) (.135 :: .135))
      (INTERCONNECT R1/A1/y R1/D4/clk
        (.150 :: .150) (.140 :: .140)))))
```

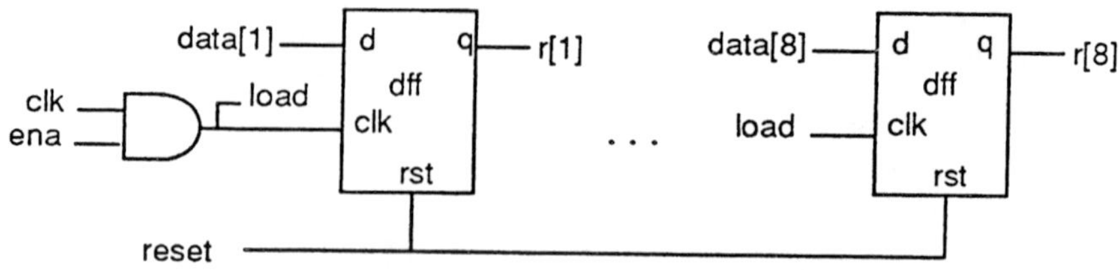

Figure 2. 4-bit Register

Standard "access routines" for VHDL

There are various methods possible for accessing delay values and applying them to generics during simulation. The next sextions describe the most feasible possibilities.

The Procedural Approach

Given the proposed standard delay file, one approach is to use a procedure for incorporating this information. The procedure body would be written in another language such as C. A possible procedure declaration would be:

```
procedure SDF_annotate
(sdf_str: string;
hier_instance: string := " ";
config_str: string := " ";
log_str: string := "sdf.log";
mtm: string := "typical");
```

Due to hierarchical references, the actual update of information would need to occur after the elaboration phase in order to have the entire design available. Using a series of access routines, the data structures corresponding to generics would be modified with the new values. The utility for annotating the values could have the following steps:

1. read the procedural arguments being passed
2. open the various files
3. read each line of the SDF file
4. compare each line to determine the type of action to perform
5. if instance, change scope to that instance

6. if type delay, locate object and replace delay
7. if timing check value, locate object and update with new value
8. repeat steps 4 thru 7 until reach end of file

As can be seen from the above example, the flow of this routine is general enough that it can be applied to any design rather than requiring each design to have its own special routine. This provides a more convenient mechanism for getting the values into the design during simulation. A big disadvantage of this method is the fact that the data structure is modified after elaboration has been performed. Additionally, a manual effort is required to update the source code to document the new values. While creating the SDF configuration file that correlates each VHDL delay name with the appropriate transition is a one time effort for each design, it will be a lengthy job, especially for large designs.

Initializing Generics

The manual effort to update the source code in the previous approach could be eliminated by some changes in the language semantics. The semantic changes would include the ability to set a generic value based on a value read from a file. For ease of use and implementation by the user, the language semantics would need to allow the name of the generic to be read from the file as well. These generic names will probably be hierarchical references in order to identify which instance's generic is to be updated.

These changes would provide the flexibility needed by the user to simplify the task of updating the design from a simulation and documentation perspective. This represents a more automated approach. However, the functionality to be performed is difficult to identify when all the timing information is imbedded in the design. To illustrate, a behavioral description of a D-Flip Flop is shown below.

```
library IEEE; use IEEE.std_logic_1164.all;
entity DFF is
  generic ( trst_hl : time := 0 ns;
        trst_lh : time := 0 ns;
        td_hl : time := 2 ns;
        td_lh : time := 1.5 ns;
        td_x : time := 2 ns);
  port ( Q: out std_ulogic;
        QB: out std_ulogic;
        D: in std_ulogic;
```

```
          CLK: in std_ulogic;
          RST: in std_ulogic);
end;
architecture arch_beh of DFF is
   signal tmp_q: std_ulogic;
begin
   process (CLK, RST)
   begin
      if RST = '1' then
         if tmp_q /= '0' then
            tmp_q <= '0' after trst_hl;
            QB <= '1' after trst_lh;
         end if;
      elsif (CLK'event and CLK = '1') then
         if (tmp_q = '0' or tmp_q = 'Z') and D = '1' then
            tmp_q <= D after td_lh;
            QB <= not D after td_hl;
         elsif (tmp_q = '1' or tmp_q = 'Z') and D = '0' then
            tmp_q <= D after td_hl;
            QB <= not D after td_lh;
         else
            tmp_q <= D after td_x;
            QB <=not D after td_x;
         end if;
      end if;
   end process;
   Q <= tmp_q;
end arch_beh;
```

Proposed Timing Relationships

Before presenting an approach that eliminates a number of the disadvantages discussed above, it is necessary to introduce a new VHDL construct. This construct allows for a representation of pin-to-pin timing delays that is more straight forward and natural for the ASIC designer. This construct consists of two statements: a RELATION declaration and a RELATION specification. The RELATION declarations could be defined in a package or in a configuration declaration; whereas, the RELATION specifications would be specified in configuration declarations. These constructs are illustrated in the following example of a D-Flip Flop.

```
library IEEE; use IEEE.std_logic_1164.all;
entity DFF is
  port ( Q: out std_ulogic;
    QB: out std_ulogic;
    D: in std_ulogic;
    CLK: in std_ulogic;
    RST: in std_ulogic);
end;
architecture arch_beh of DFF is
begin
  process (CLK, RST)
  begin
    if RST = '1' then
      Q <= '0';
      QB <= '1';
    elsif (CLK'event and CLK = '1') then
      Q <= D;
      QB <= not D;
    end if;
  end process;
end arch_beh;

configuration technology_1 of DFF is
  for arch_beh
    -- relation declarations follow
    relation pin_delay : time of (P1, P2 : port);
    relation setup_check : time of (D, CLK : port);
```

```
    relation cap : capacitance of (P : port);
    relation intrinsic_delay : time of (Pin, Pout : port);
    relation slope : capacitance of (Pin, Pout : port);
    relation cell_area : real of (A : architecture);
    -- relation specifications follow
    for pin_delay of D to Q use propagation_delay
      ('0' -> '1' =>6 ns, '1' -> '0' =>8 ns);
    for pin_delay of RST to Q use propagation_delay
      ( all => 0 ns);
    for pin_delay of all to QB use propagation_delay
      (all => 0 ns: 1 ns: 2 ns);
    -- min. : typ : max delay values
    for setup_check of D to CLK use time_check
      ( '0' -> '1' => 10 ns);
    -- the '0' -> '1' corresponds to the trigger signal
    -- i.e clk
    for cap of D use ( 9.0 pf);
  end for;
end technology_1;
```

As seen from this example, the functionality of this device is described in the architecture, while the technology-specific information is given in the RELATION constructs. The above relation constructs allow the user the flexibility to choose the relation names. In the relation specifications, a key word such as `propagation_delay` or `time_check` identifies the type of action to be executed during simulation. The actions to be performed are implicitly done. Alternatively, a standard timing package containing relation declarations could be provided where the relation name indicates a specific action to be done. This option would limit the user to those relations defined in the package.

Suppose `Q = '0'` and `D = '1'` when a rising edge occurs on `CLK` in the above example. The propagation delay relation, `pin_delay of D to Q`, is evaluated when the statement `Q <= D;` is executed in the above example. The new value of `Q` would be scheduled to take effect in 6 ns. Likewise, if the value of `Q` was to change from `'1'` to `'0'`, then this value change would be scheduled to occur 8 ns from now. For a particular transition, delays representing minimum, typical, and maximum times may be specified. If only one value is specified for a transition, then this value is used to satisfy each of the three delay times.

During simulation, timing check relations are evaluated whenever the data and trigger signals change. In the above setup_check example, when a `'0'` to `'1'` transition is detected on `CLK`, `D'last_event` is implicitly compared to the specified time. If this value is less than the specified time, then an error message is reported.

With the availability of relation constructs, a utility would read the SDF and create a configuration file that contains the corresponding relation statements. There are several advantages to this approach. First, it can be done as a step separate from simulation rather than invoking a procedure to modify the data structure. Second, the new values being used in a simulation run are clearly documented. Third, only analysis of the configuration file is required before proceeding with simulation. Another key advantage is the separation of functionality from technology specific data, which allows the user to perform more efficient trade-off analysis.

Besides the above advantages, synthesis and simulation tools can use the same design—thus providing consistent information. This reduces the number of libraries an ASIC vendor needs to maintain for simulation and synthesis purposes. While some of the information may be applicable to only one of the tools, the complete design is clearly documented.

Conclusions

Language enhancements related to timing are necessary for two reasons: first, to update a design based on data derived from other tools, and 2. to have a more direct representation for VHDL timing constructs. These enhancements include allowing use of hierarchical names, initializing generics based on reading generic names and their corresponding values from a file, and adding a new construct such as the RELATION construct. These VHDL issues would be addressed by the 1992 Design Requirements Group.

The SDF format is useful for inter-tool communication. The CAD Frameworks Initiative (CFI) would address issues related to the SDF.

Summary

VHDL was designed to provide a standard notation for electronic designs. In general it has proven adequate for this purpose. Timing is one of the few areas not realized adequately for ASIC designs. This paper has shown how some of these inadequacies can be addressed.

References

[1[IEEE Standard VHDL Reference Manual, IEEE Std 1076-1987, The Institute of Electrical and Electronic Engineers, Inc., New York, NY, March 31, 1988.

[2] Cadence Standard Delay File (SDF) Format Specification, Version 1.1 Preliminary, Cadence Design Systems, Inc., Lowell, MA, May 31, 1991.

Part II

SYNTHESIS

A VHDL-Driven Synthesis Environment

Haluk Konuk[1] F. Erich Marschner

CAD Language Systems, Inc.
15245 Shady Grove RD, Suite 310
Rockville, MD 20850, U.S.A.

Tel: (1) 301 963 5200
Fax: (1) 301 963 1511

1. INTRODUCTION AND PROBLEM DESCRIPTION

As the complexity and the size of the VLSI systems grow steadily, the complexity a human mind can grasp remains the same. This fact forces designers to use design automation tools. It is almost impossible for a designer to visualize the gate-level structure of a reasonably complex VLSI chip, not to mention its transistor-level structure. Automatic placement and routing tools have already become an integral part of VLSI design. Synthesis tools are now on the rise to help designers cope with complexity at higher levels.

Logic synthesis tools may perform several functions to help the hardware designer explore the design space in a short time. They may apply optimizing transformations to an existing logic circuit to improve the area or the time performance of the circuit, or they may automatically generate a logic circuit from the state transition table of a finite state machine or from the logic behavior specified in a hardware description language, such as VHDL. Logic synthesis research is conducted by several institutions, and has already resulted in various commercial logic synthesis tools. The need and the demand for such tools are growing each day.

---

[1] Mr. Konuk's current address is: Department of Computer Engineering; University of California, Santa Cruz; Santa Cruz, CA 95064

J. Mermet (ed.), VHDL for Simulation, Synthesis and Formal Proofs of Hardware, 101–115.

High-level synthesis tools produce a register-transfer level structure (data path) and a sequence of operations to be performed on the data path [McPC88]. The data path is a network of functional units, storage units, and interconnection units. The input to a high-level synthesis tool is an abstract behavior, usually represented as an algorithm described with standard programming constructs, such as loops and conditional statements.

The arithmetic and logic operations given in the input algorithm are scheduled into specific control steps by the high-level synthesis tool, and this schedule determines the sequence of operations to be performed on the data path. The logic circuitry (control path) that will emit the necessary control signals to the data path to implement this sequence may be generated by a logic synthesis tool.

High-level synthesis research has gained a big momentum during the recent years. Commercial products are about to appear in the market in addition to already existing in-house tools. High-level synthesis tools are inevitable for design space exploration at the system level.

VHDL is emerging as the global standard for digital design specification for all stages of hardware design. VHDL has the power to specify the behavior of a system using a mixture of behavioral, data flow, and structural levels. The language can be used to describe the specification of inputs as well as the synthesis results for both logic synthesis and high-level synthesis applications.

A digital system specification usually involves different levels of abstraction. It is not practical to express all designs, or even any single design, in purely algorithmic form, or purely data flow form, or purely structural form. In a fully-automated design environment, high-level synthesis tools together with logic synthesis tools and module generators will need to work cooperatively to process different portions of the same design.

Therefore, there is a need for a common representation format that can be used to represent the system specification, the partial design data generated by various synthesis tools, and the synthesis results produced by those cooperating tools. This representation format serves as a common foundation that constitutes a synthesis environment in which a suite of synthesis tools can be brought to bear on various portions of the same design. An input design specification written in VHDL may be mapped onto this foundation, and similarly the synthesis results may be mapped onto VHDL from this foundation.

[DuHG90] is a recent attempt to develop an intermediate representation for synthesis. However, it is intended only for behavioral synthesis. There is a need for a more comprehensive approach that deals with algorithmic behavior, with data flow, with timing constraints, with pre-existing structure, and with both the behavioral hierarchy and the structural hierarchy.

This paper presents our research in developing a synthesis representation format that serves as a foundation for various design synthesis and analysis applications. New tools can be developed or existing ones can be integrated based on this foundation which also has a VHDL front end. The block diagram of the complete design environment we envision is illustrated by Figure 1.

The mixed-level VHDL specifying the design is analyzed by a VHDL analyzer that creates a VHDL View library unit within the Design Library System. The VHDL View library unit is then processed by the Synthesis Preprocessor to create a Synthesis View library unit within the same Design Library System. The Synthesis

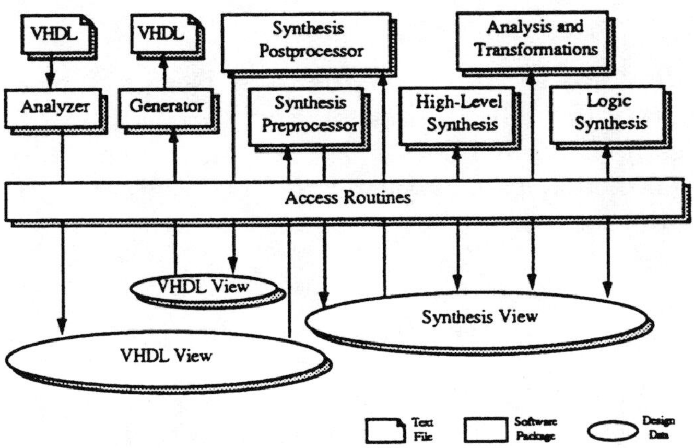

Figure 1. A Synthesis-Oriented VHDL Design Environment.

Preprocessor populates only the design specification portion of the Synthesis View library unit; the other portions will be populated by a suite of synthesis tools.

This suite may contain high-level and logic synthesis tools, even if these tools were not designed to cooperate. Analysis and transformation tools may be a part of a synthesis tool or may control the invocation of synthesis tools. They may perform the analysis of the partial synthesis results to make synthesis decisions. They may perform high level transformations, such as unwinding loops, common subexpression elimination, merging or creation of processes, etc.

One major contribution of this paper is the usage of the same representation mechanism, namely control-data flow graphs, to represent the design data for both high-level synthesis and logic synthesis applications. This allows the user to experiment both high-level synthesis and logic synthesis tools on the same portion of the design.

Synthesis tools are allowed to make incremental refinements to the data structures in a Synthesis View library unit towards the concrete realization of the design. The synthesis activities terminate when a satisfactory design represented by a netlist of primitive hardware resources drawn from a cell library is constructed. The Synthesis Postprocessor reads the synthesis results from Synthesis View library units and creates a VHDL View library unit that will be mapped to structural VHDL by a VHDL Generator.

The following section describes the Design Library System. Section 3 describes the representation capabilities of Synthesis View in detail. Section 4 uses a VHDL example, and illustrates the synthesis tasks that can be performed on Synthesis View library units. Finally, section 5 is summary and status.

2. DESIGN LIBRARY SYSTEM (DLS)

Our Design Library System (DLS) is an object-oriented data base. A library unit within the DLS is a collection of various types of objects linked with pointers to form a meaningful structure. Each object can point to other objects via its attributes. A set of object definitions form a specific View. The View type of a library unit is determined by the set of object definitions whose instances are used to populate the library unit. For instance, the set of object definitions for representing concepts and stages of the synthesis process are used to create Synthesis View library units which contain the instances of those object definitions.

Our system includes a set of C routines which can be called by any software tool to acces DLS library units. These access routines, as shown in Figure 1, form an interface (Software Procedural Interface - SPI) between the design data and the software tools. SPI provides extensive error checks while servicing design data access requests.

3. EXTENSION OF THE DLS TO SUPPORT SYNTHESIS

Synthesis View is a set of object definitions whose instances can be structured to represent all stages of synthesis processes. Synthesis View has three major portions. These portions are grouped under StateMachine, LogicNetwork, and BlackBox object definitions. Thus the Synthesis View is effectively a representation medium for modeling the synthesis process.

Manipulation of the data structure under a StateMachine object is primarily the concern of a high-level synthesis process. Similarly, manipulation of the data structure under a LogicNetwork object is primarily the concern of a logic synthesis process. A BlackBox object represents a portion of a design that does not need to be synthesized, since it may have been pre-synthesized or manually designed. A Synthesis View library unit may even contain only BlackBox objects in order to describe the interconnection of the synthesized parts.

The fundamental mechanism for representing behavior in the Synthesis View is the control-data flow graph (CDFG). A CDFG represents the control and data operations that are performed within a portion of a design as well as the values that are produced and consumed by those operations. Before we go into discussion of the three major portions of the Synthesis View, we will first describe the object definitions used to represent CDFGs and timing constraints.

3.1 Representation of Hierarchy of CDFGs with Timing Constraints

A CDFG is modeled by a LocalRegion object. A LocalRegion object corresponds to either a VHDL process statement, or a loop statement, or a subprogram body, or a single branch of a conditional statement, or a set of concurrent signal assignments. The control flow in these VHDL constructs is represented by explicit control operations, such as a reiterate operation to reiterate a loop body, a terminate operation to terminate a loop or a subprogram body, fork-join operations for conditional statements, etc.

Nodes of CDFGs are modeled by SimpleOperation and Channel objects. A SimpleOperation may represent either a data operation or a control operation, such as invoking another CDFG. A Channel may represent a single value produced by an operation node of a CDFG, or a constant value, or an input to a LocalRegion, or a VHDL signal object that provides communication among Partitions (a Partition is either a StateMachine, or a LogicNetwork, or a BlackBox). A Channel object has a bit structure derived from the VHDL types and subtypes for the corresponding VHDL variables and signals. The arcs of the CDFGs show the flow of data.

A TimingConstraint object specifies a constraint imposed on a) the time needed to execute all the operations of a LocalRegion, and b) the time interval between two CDFG operations in a LocalRegion. Time expressions of VHDL **wait for** statements and **after** clauses are mapped onto TimingConstraints by the Synthesis Preprocessor. Additional TimingConstraints can be entered to the Synthesis View via attached tools.

3.2 StateMachines

A StateMachine object contains:

i) A hierarchy of CDFGs with timing constraints. The CDFGs and their hierarchy are derived from the input VHDL process statements and concurrent procedure calls. The Synthesis Preprocessor performs this derivation.

ii) The scheduling of the CDFG operations into control steps. One of the two major tasks of a high-level synthesis system is scheduling. The schedule of operations has a very significant effect on the size of the data path and the total time needed by the data path to execute all the operations.

iii) The allocation of technology-independent register-transfer level (RTL) resources and their interconnection (the data path). The other major task of a high-level synthesis system is allocation. The allocation of resources determines the size of the data path and sometimes the length of the schedule if allocation is done before scheduling.

iv) The binding of RTL resources to the CDFG operations and values. This information forms the specification for the control path, which will most likely be synthesized by a logic synthesis tool.

v) The binding of RTL resources to technology-dependent library components. This link is established by pointing to a Symbol for the component. This Symbol may be the name of a LogicNetwork which contains the logic circuitry made up of gates from a certain technology, or may be the name of a component from a cell library.

vi) The binding of CDFG operations to VHDL operators and the binding of CDFG values to VHDL variables. This information allows the designer to keep track of the steps synthesis tools take all the way from VHDL to CDFGs, to the schedule, and to the data path.

The representation capabilities listed above under a StateMachine object allow us to store the design data representing the system specification, and any increment of information generated by synthesis tools, such as partial scheduling or partial allocation, or partial binding, or complete synthesis results.

3.3 LogicNetworks

A LogicNetwork object contains:

i) A hierarchy of CDFGs with timing constraints. These CDFGs represent the logic behavior and they are derived from sets of

VHDL concurrent signal assignment statements by the Synthesis Preprocessor.

ii) The allocation and binding of technology-independent logic components. Each CDFG data operation such as, an AND, an OR, etc., corresponds to a technology-independent logic component. Since there is no scheduling of CDFG operations, a logic component can be bound to only a single CDFG operation. Therefore, allocation and binding of technology-independent logic components is straightforward, and a logic synthesis tool can apply its area-time optimizing transformations directly to the CDFGs.

iii) The binding of logic components to technology-dependent library components.

iv) The binding of CDFG operations to VHDL operators and the binding of CDFG Channel objects to VHDL signals.

The representation capabilities listed above under a LogicNetwork object allow us to store the design data which contain system specification, and any increment of information generated by logic synthesis tools.

Note that the information items a LogicNetwork object contains is a subset of the information items contained by a StateMachine object. This is because we use the same mechanism of design data representation for both high-level synthesis and logic synthesis applications, and a logic synthesis process does not involve a scheduling task.

3.4 BlackBoxes

A BlackBox object is created from each VHDL component instantiation statement by the Synthesis Preprocessor. A BlackBox does not need processing by a synthesis tool. It may point to the name of an object in any library unit that contains the behavior of the BlackBox and perhaps also contain the synthesized or manually designed structure for that behavior.

4. DEMONSTRATION OF SYNTHESIS VIEW USAGE

This section includes a mixed-level VHDL example that will be used to demonstrate the usage of Synthesis View as a common foundation for various high-level synthesis and logic synthesis applications. Following is the VHDL example.

```
ENTITY e IS
    PORT (A, B, C, D : IN  word;
                   Z : OUT word);
END e;


ARCHITECTURE a OF e IS
    SIGNAL S, T, F, G: word;
    COMPONENT box
      PORT (S: IN  word;
            T: OUT word);
    END COMPONENT;
BEGIN

P1: PROCESS (A,B,H)
        VARIABLE x,y: word;
    BEGIN
        x := A and B;
        S <= x AFTER 300 ns;
        y := x + B - H;
        S <= y AFTER 500 ns;
    END PROCESS;

P2: PROCESS (C)
        VARIABLE i, product: word;
    BEGIN
        i := C;
       product := 1;
        FOR j IN 1 TO 3 LOOP
            product := product * i;
            i := i + 1;
        END LOOP;
        H <= product;
    END PROCESS;

    F <= A nand D;
    G <= not (A and D and B);
    Z <= not (F and G and T);

    B1: box PORT MAP (S, T);

END a;
```

The Synthesis Preprocessor creates two StateMachine objects for processes P1 and P2, a LogicNetwork object for the three concurrent signal assignment statements, and a BlackBox object for the component instantiation statement. It also captures the interconnections among those partitions. Figure 2 illustrates the top-level structure of the SynthesisView derived from the example.

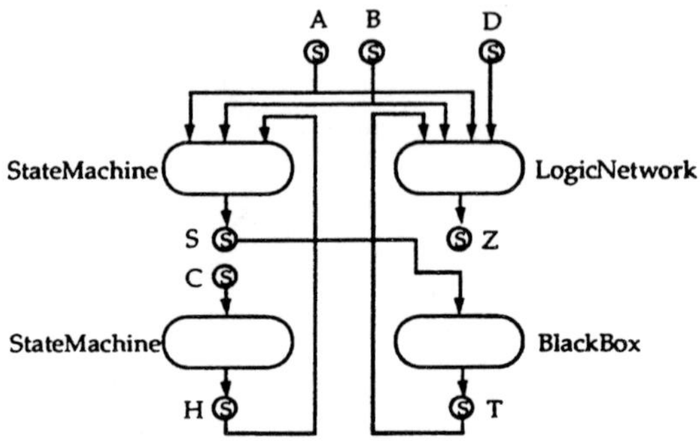

Figure 2. Top-level Structure of SynthesisView Derived from Example.

The StateMachine object for P1 initially contains only the CDFG with timing constraints for the corresponding process statement as shown in Figure 3. A high-level synthesis tool processes this StateMachine object.

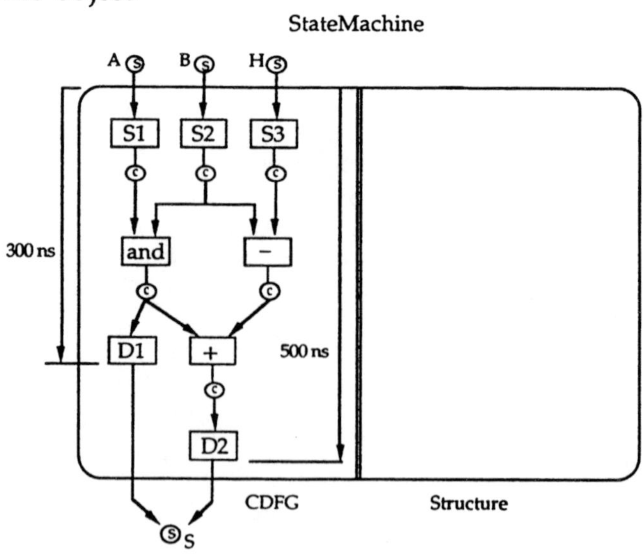

Figure 3. Initial Contents of StateMachine Object for Process P1.

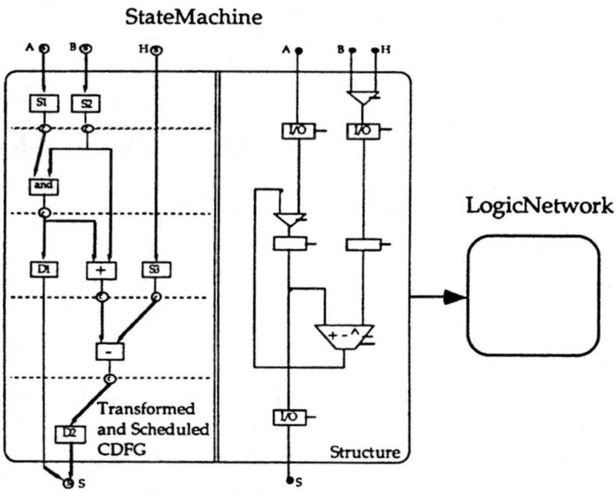

Figure 4. Results of Synthesizing the StateMachine Object for Process P1.

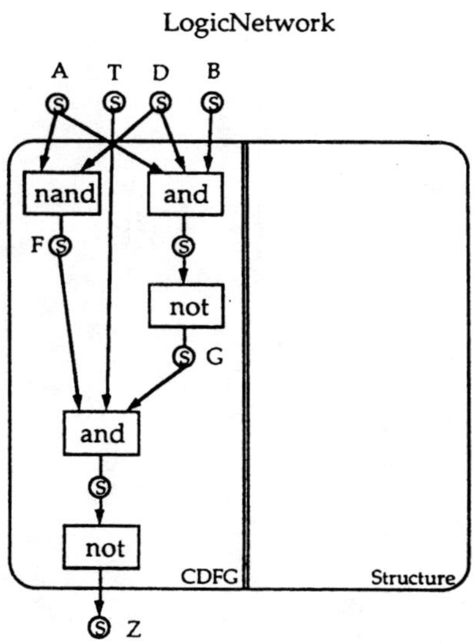

Figure 5. Initial Contents of the LogicNetwork Object from Example.

The synthesis process involves tranforming the CDFG by swapping the order of addition and subtraction (which will yield a smaller data path), scheduling the CDFG, and allocating the RTL resources, as shown in Figure 4. A clock cycle time of 100ns is assumed. Binding of RTL resources to the CDFG is not shown in Figure 4 for simplicity. The StateMachine in Figure 4 points to a new LogicNetwork object which contains the implementation of the conrol path synthesized by a logic synthesis tool.

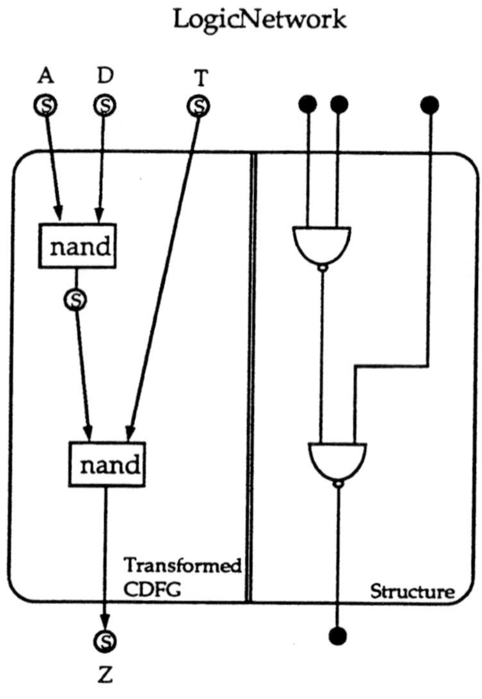

Figure 6. Results of Synthesizing the LogicNetwork Object.

The LogicNetwork object in Figure 5 initially contains only the data-flow graph for the three concurrent signal assignment statements. A logic synthesis tool applies a boolean transformation, and creates the results shown in Figure 6. Binding of logic components to the data-flow graph is obvious. This transformation is taken from [DBGJ84].

The Synthesis Preprocessor creates a StateMachine object for process P2 to be processed by a high-level synthesis tool. But, an analysis tool may decide to use logic synthesis instead; to do so, it transforms the original StateMachine to a LogicNetwork by unwinding the **for** loop, as shown in Figure 7. This is only one example of behavioral transformations. Several others, such as in-line expansion, process creation or merging, common subexpression elimination, grouping of operations, etc., can be applied to the Synthesis View.

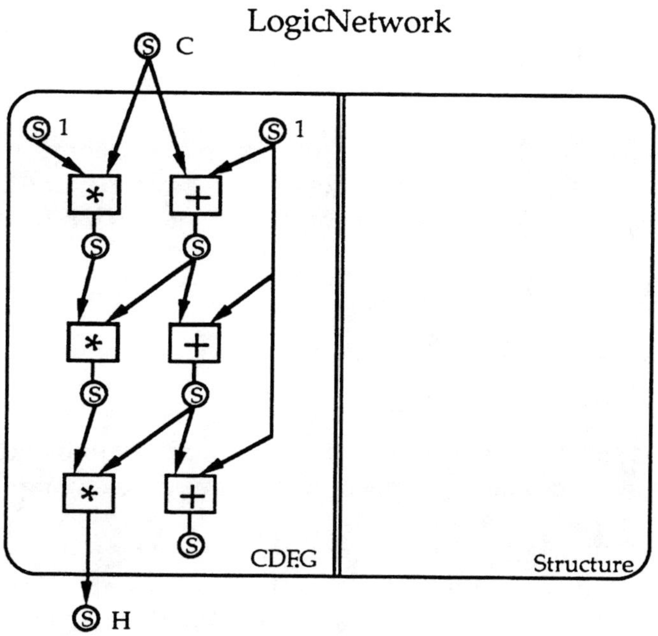

Figure 7. LogicNetwork Object Resulting from Behavioral Transformations.

5. SUMMARY AND STATUS

In this paper, we presented a comprehensive design data representation scheme for a unified synthesis environment with a

VHDL front-end. This representation scheme is called Synthesis View. Following is a summary of the Synthesis View features:

- Can represent both system specification and synthesis results.
- Supports hierarchy both in behavior and structure.
- Supports mixed-level VHDL.
- Supports integration of a suite of synthesis tools that will work on different portions of the same design.
- Can represent the increments of a synthesis process (intermediate synthesis results).
- Uses CDFGs to represent both the algorithmic and logic behavior, which enables the designer to experiment both high-level and logic synthesis tools on the same portion of the design.
- Supports behavioral transformations.
- Provides a complete linkage between the behavior and the structure.
- Can be used as a research vehicle to experiment and compare various synthesis algorithms, ans as a design exchange format.

The VHDL example used in section 4 served to demonstrate some aspects of Synthesis View, but many other examples with more detailed drawings are needed to illustrate all the aspects of Synthesis View. The set of object definitions for Synthesis View is complete, but it will certainly be evolving as time passes. The VHDL Analyzer, VHDL Generator, and Synthesis Preprocessor shown in Figure 1 are completed, and work is progressing on various analysis, transformation, and synthesis elements of the system.

REFERENCES

[McPC88] M.C. McFarland, S.J., A.C. Parker, R. Camposano. Tutorial on high-level synthesis. Proc. DAC 1988.

[DuHG] N.D. Dutt, T. Hadley, D.D. Gajski. An Intermediate Representation for Behavioral Synthesis. Proc. DAC 1990.

[DBGJ84] J.A. Darringer and et al. LSS: A System for Production Logic Synthesis. IBM J. Res. Develop. Vol. 28 No. 5 Sept. 1984.

VHDL Specific Issues in High Level Synthesis

Adam Postula
Swedish Institute of Microelectronics
Isafjordsg. 22, 164 21 KISTA, SWEDEN
email : adam@inmic.se

Abstract

VHDL is essentially a simulation language which is now tried to be used throughout the design process, not only for simulation but also for synthesis and test. High Level Synthesis methodology fits well to the idea of using a specification/simulation language like VHDL for the design of digital circuits. Unfortunately, obvious advantages of using one hardware language in the design process can be outweighed by the language incompatibility with some of the design tasks; incompatibility from the user as well as from the tool developer point of view. This paper analyses some essential features of VHDL as the synthesis language, points to problems and suggests practical solutions which do not demand changes in VHDL 1076 standard. Some questions of efficiency of the synthesis process are also raised that lead to requirements for changes or completions of present VHDL.

VHDL in "design by simulation and synthesis" methodology

Modern electronic circuit design process can not be envisaged without CAD design tools based on simulation and synthesis. The design methodology based on simulation and synthesis can be briefly described as a loop starting with a behavioural design description for simulation. This model can hide all the implementation aspects and can also be considered as a purely behavioural specification of the design. Once simulated and accepted this description is further partitioned, more and more details are added and the result resimulated. Finally, such a level of details in the model is reached that the synthesis system can take over further elaboration and produce automatically a hardware implementation of the described behaviour. That implementation can be considered as correct by construction or verified functionally by simulation or by formal verification. Fig.1 depicts the design process based on simulation and synthesis.

The fundamental assumption here is that *the synthesis system produces hardware which behaves correctly and exactly like purely behavioural models simulated earlier.* Any deviation from this principle should be treated with suspicion since it contradicts the assumed method: first create and simulate models then get them automatically designed. As we will see in next sections of

J. Mermet (ed.), VHDL for Simulation, Synthesis and Formal Proofs of Hardware, 117–133.

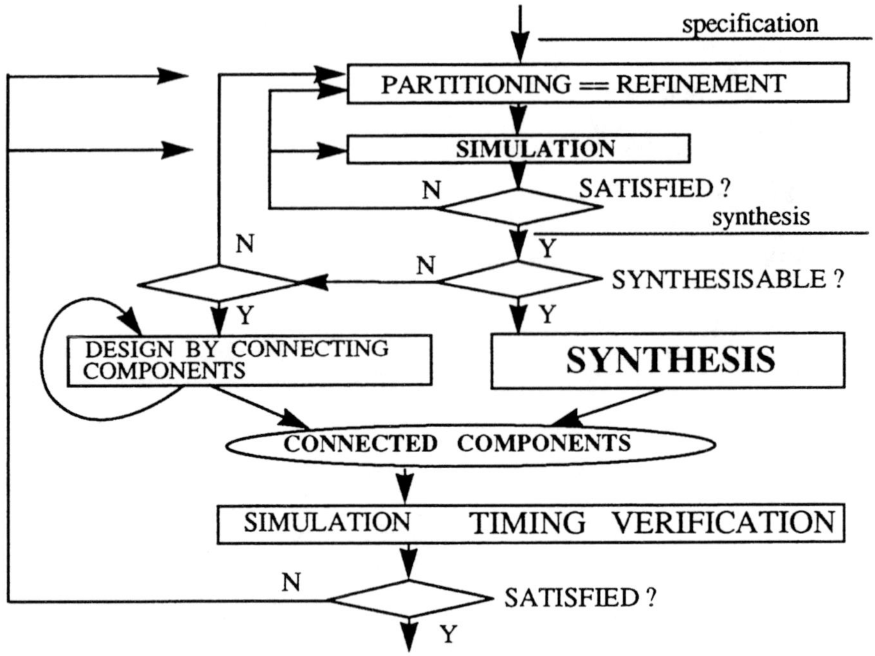

Fig. 1 Interaction simulation - synthesis

this paper there are some intrinsic problems with VHDL regarding this fundamental requirement. Those problems are mainly related to timing and depend on the lack of clearly defined synthesis semantics of the language.

Another premise, more of practical importance, is that *even when describing behaviour for simulation, the designer works with synthesis in mind (at least at later stages of the design process).* This is unavoidable in any design process with any tools but can result in that the designer knowing limitations of a particular synthesis system makes descriptions which suite one synthesis system better than another. VHDL gives the designer a very good support regarding possibility of making different views of the same design entity with different architectures. Functionally equivalent but tool specific descriptions can be easily made and kept coherent in VHDL environment.

One more practical requirement of the methodology is *partitioning the design into "reasonable" chunks of behaviour interacting with each other.* Partitioning principle is commonly applied by engineers to all their designs in order to handle complexity. In high level synthesis it increases the chances of synthesising a good solution because the synthesis system works with smaller tasks. Partitioning on the other hand must be taken with care because it can lead to too low level of abstraction in the description, even though much higher level can be handled by synthesis tools. A typical mistake of a designer is to describe his design as FSMs and try to synthesise that with architectural synthesis tools. VHDL supports partitioning both in the behaviour and in structural descriptions.

The last but not least aspect of methodology is something not seen in fig. 1 - an information on the target technology which makes partitionings of the descriptions efficient and an average human made design so performance/cost effective. The requirement for the synthesis can be formulated as : *the technology information must be available and used by the synthesis system as early as possible in the synthesis process.* The behavioural description taken as input by the synthesis system is technology independent but the synthesis process itself must be very much aware of the target technology when applying optimisations. This will assure effective hardware solutions. VHDL can be used to build generic library of functional modules which stores technology information in high level synthesis. VHDL models of these modules allow simulation of synthesis results for purpose of verification.
As we have seen from the above analysis of methodology requirements, VHDL as a description language suits well the described methodology. It allows data type abstraction, partitioning of the description, purely behavioural models interacting with each other. Behavioural abstract data type models can be mixed with lower abstraction models and even with the synthesised hardware. What VHDL lacks is clearly defined synthesis semantics. The synthesis system must generate hardware behaving exactly like the simulated model, in other words the simulation semantics of VHDL must be preserved in the generated hardware. Is this what we want when designing highly optimised hardware ? Well, it is may be acceptable until it does not cause too much inefficiency in the hardware solution. The ideal design language should of course have both simulation and synthesis semantics coherent and allowing efficient hardware synthesis. In the subsequent sections of this paper we will analyse VHDL from the synthesis point of view and try to find how efficient translations from VHDL behavioural descriptions to hardware can be made and what problems must be solved, both in the coming modifications of the language and in HLS tools.

VHDL as a language for synthesis.

High Level Synthesis in short is a translation of a program-like behavioural specification to a set and connection of functional hardware modules such that the specified behaviour is preserved and the implementation is optimal according to some criteria. There are several synthesis systems which accept VHDL, or more correctly a subset of it as input language [1],[2],[3]. The common strategy for using VHDL in these systems is describing the design as a set of communicating processes, which are synthesised one at a time. Processes are synthesised as separate hardware modules composed of a single FSM controller and a data path. Hardware can not be shared among processes. The synthesised hardware module is synchronous though different modules can have different clocks. Usually, no predefined synchronisation between processes is supported by the synthesis system, the designer must describe it explicitly. One of representative subsets used for synthesis is SynVHDL used at Swedish Institute of Microelectronics in the SYNT system [4]. We present briefly SynVHDL in order to give the reader an overview of all the constructs used for synthesis. In the next few sections we will analyse those constructs which cause problems in representation and in interpretation for synthesis. Sub-setting of VHDL stems from the synthesis strategy described earlier.

SynVHDL structural descriptions and packages

SynVHDL allows structure of the design to be described as interconnection of components which are separately synthesised or marked as not synthesisable.

SynVHDL allows:

Packages with declaration of user data types.
Entity with port declaration only.
Architecture with declaration of data types, one process in the architecture body.
Process with data type declarations, variable and constant declarations.

Structural hierarchy in the design and usage of standard components is supported by component declaration and component instantiation statements, which are recognised as not synthesisable by SYNT and ignored, resulting in unchanged interconnections of components. Abstract data types in that interconnection are translated then into the synthesised data types and a final structural VHDL description is made.

SynVHDL data types and attributes.

SynVHDL allows at present :

BOOLEAN
BIT
INTEGER And its subtypes.
BIT_VECTOR and its subtypes.
ARRAY OF INTEGER
ENUMERATION.

Array of integer is mapped to a memory module and at present can be declared only for use in variable (RAM) and constant declarations (ROM) within a process. The implemented memory is a part of the data path of the module the process is synthesized to, which means it can not be shared by other processes (a temporary limitation of SYNT).
Presently defined attributes for synthesis purposes are following :
clock, reset, ram, rom, pin_number, module(n), chip(n).
The clock attribute can be set to indicate that the signal works as a clock for the specified process. Only one input signal to each process can be attributed as clock. This information is used by the system when no sensitivity list is present on WAIT statements used in the process. The signal attributed as clock is assumed there.
The reset attribute can be set on a signal to indicate that this signal is used for desribing the reset behaviour of the hardware. The name of this signal will be used by SYNT for the synthesised reset signal it generates. Since SYNT has a strategy of generating reset equivelent to restarting the simulator, the designer must assure that the behaviour of a reset attributed signal is conformant with that. The use of this attribute is when the designer needs to simulate reset actions in the model but wants to avoid an inefficient implementation of reset (see section VHDL process description inefficiencies).

The pin_number, module(n), chip(n) attributes are used for specification of physical partitioning of the design. They can be set before or after the synthesis is done.

SynVHDL operators

Arithmetic on INTEGER :	+, -, *, sign.
Relational on INTEGER, BIT, BIT_VECTOR:	=, /=, <<, >, <<=, >=.
Logical on BOOLEAN, BIT and BIT_VECTOR:	AND, OR, XOR, NOT.

Division is not supported yet depending on availability of functional modules in our library of generic components. Concatenation on bit_vectors will be supported soon.

SynVHDL process control and synchronisation.

SynVHDL allows following sequential control constructs :

IF THEN ELSIF .. ELSE
CASE
WHILE LOOP
FOR LOOP.

Any combination and hierarchy of the above constructs is accepted. No general LOOP with EXIT and NEXT is allowed since in most cases WHILE and FOR loops are cleaner solutions for the designer.
Synchronisation with other processes (sources of incoming signals and receivers of the generated signals) is handled by WAIT statements.
Allowed are :
WAIT ON clock UNTIL (clock = 'value') AND condition_on_incoming_signals ;
WAIT FOR n*clock_period ;
Clock or clock attributed signal as the only element on the WAIT sensitivity list assures that the simulated behaviour corresponds to that of the synchronous synthesised module.
If the list is left empty the system assumes a clock attributed signal on it and gives the designer a message about possible discrepancy in the behaviour.
WAIT FOR n*clock_period is interpreted as a timing synchronisation statement controlled by clock.
At present SynVHDL does not allow AFTER clause for timing specification. This is due to the synchronous design principle and fussiness in AFTER time specification (max , min or something else ?) which will be discussed later.

Signals

Signals can be of type : BOOLEAN, BIT, ENUMERATION, INTEGER and their subtypes, BIT_VECTOR and its subtypes.

Array types are not supported yet on signals (this would need some predefined communication with a separate memory module if such a signal is read from many processes).

No resolved signals are allowed yet. VHDL semantics makes resolution of signals transparent to involved sources by demanding execution of resolution functions in zero simulation time. Supporting VHDL resolution mechanism would in general demand a signal (bus) arbitration unit. Usually such an arbitration can not be done within one clock period and even less probable in 0 time as specified by VHDL simulation semantics. Synthesis can support easily resolution on "hardware close" types as bit and bit vector for such resolution schemes as open collector or tri-state bus, such resolutions will be supported by the SYNT system.
General resolution requiring bus arbitration is handled best by the designer himself specifying the behaviour of the arbiter in a separate process, communicating with signal sources. SYNT will support in future some predefined arbitration protocols on resolved signals but by the very nature of a predefined protocol the resolution function must also be predefined in the system and the designer can use that if it suites his needs. A typical example can be a predefined protocol for writing to and reading from a common memory unit.

Supported modes in port declaration : IN, OUT.
INOUT mode is not supported yet. It does not pose implementation problems within the synthesised process, the problem is its most probable use as a port connected to a resolved signal. INOUT can also be a cause of an inefficient hardware solution if used by the designer only to be able to read the value of a signal generated within the synthesised architecture (as required by the simulation semantics of VHDL).

Variables

All supported types can be used in variable declarations. Variables of array type define RAM modules (only within a processes, and each declared variable maps to a separate memory model at present).

Constants

All supported types can be used in constant declarations. Constants of array type define ROM modules (only within a processes, and each declared constant maps to a separate memory at present).

Functions and procedures

Functions and procedures are not allowed yet in SynVHDL. Following problems must be considered and solved before we can accept them:

1. Hierarchy of functions and procedures if preserved in the generated hardware can result in very inefficient solutions. Making decisions about expanding functions and flattening the hierarchy is difficult and the only reasonable solution is to rely on the designer in this task. The designer should have some support in doing expansions interactively.
2. Functions and procedures if NOT declared within a process can result in sequential modules separate to the process (processes) they are called from.

A proper communication protocol must be defined assuring that the function(-procedure)-FP module is not called by more than one process at a time or some FPmodules must be repeated and "inserted" into processes.
Again, may be, only the designer can make reasonable decisions.

3. Concurrent procedure calls are equivalent to processes with sensitivity list consisting of input parameters to the procedure. In many cases it would be more clear to define a process performing the same function.
SYNT system will support soon flattening of functions declared within the synthesised process.

VHDL representation for synthesis

VHDL representation for synthesis does not differ essentially from representation of any other hardware description language. Such a representation must reflect and preserve both the data dependency and control sequence in the VHDL process. The difficulty lies in devising an efficient representation which is easy to use by various modules of the synthesis chain and in defining translation rules which preserve VHDL simulation semantics. A graph based representation called VSG, for VHDL Synthesis Graph, is used in our system and will serve here as a media for explaining various situations in VHDL which must be properly represented in order to preserve the simulation semantics. VSG is a common design representation for all the tools in the SYNT synthesis chain and therefore it holds both behavioural information derived from the VHDL process and information about the synthesized hardware successively added by modules of the synthesis chain. Fig. 2 shows graphical representation of operator and join nodes.

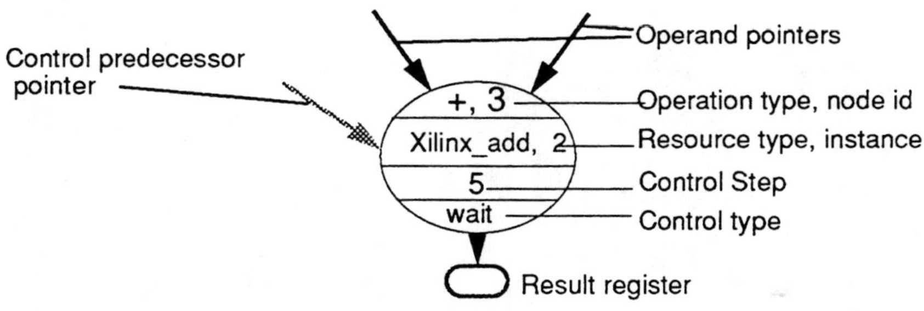

Fig. 2a VSG operation node.

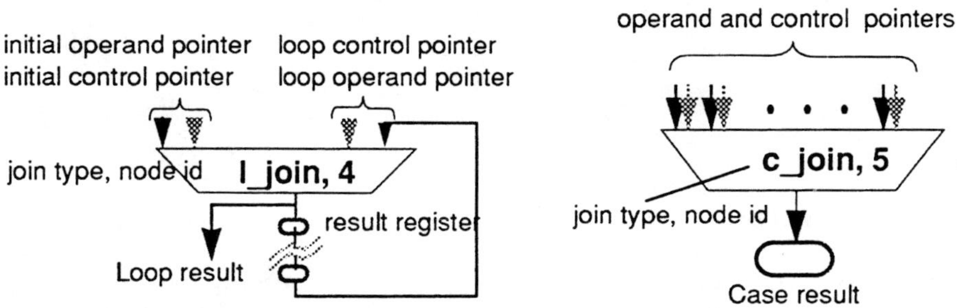

Fig. 2b VSG join nodes.

VSG has two types of nodes :
operator nodes - VHDL operators and SYNT predefined operators. This nodes are mapped to hardware primitives supporting the operator,
join nodes - gathering the data generated in different branches of control block as LOOP, CASE, IF THEN ELSE. This nodes are mapped to multiplexers, though mapping is not one to one because of multiplexer optimisation in the synthesis chain. Multiplexers can also be converted to buses when appropriate.
There are data and control arcs in VSG graph. A data arc connects the source of a data with its recipient, it gives information about data dependency between nodes. A control arc points from a node to the conditional node of a control block, giving information on how a particular node is dependent on the control flow. Each node has a number of fields storing information about e.g.: operator type, a pointer to its control predecessor, the hardware unit the operator is executed on (filled by the allocator and operation to module binder). The operand field holds information about operand width, bound multiplexer and bound multiplexer input. The result field stores information about its width and the bound register. Join nodes have a number of operands (data arcs) delivered by different branches of loop, case, if blocks. Every operand of a join node has its corresponding control arc pointing to the conditional of the respective control block. All the hardware fields are place-holders only, it depends on optimisations if they are filled or bypassed.
Translation of most control constructs in a VHDL process is fairly straightforward. As an example an IF THEN ELSE and WHILE LOOP representation is shown in fig.3.

```
 c := val ;
 IF (e = '1')
 THEN
    WHILE c < 10  LOOP
       c := c + 1 ;
    END  LOOP ;
 ELSE
    c := 0 ;
 END IF ;
```

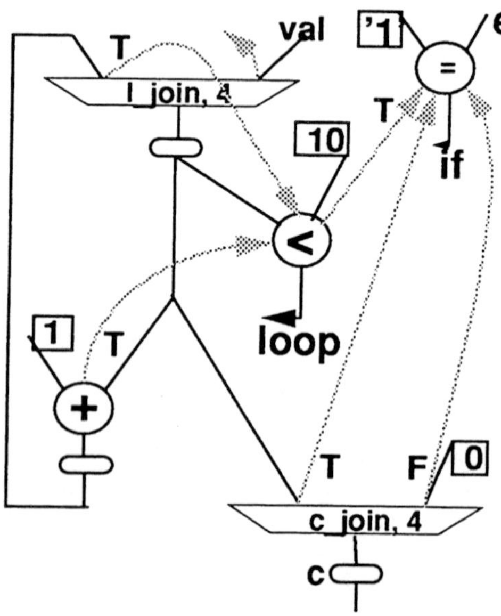

Fig. 3 VSG representation of IF THEN ELSE and WHILE LOOP

Any control block starts with its conditional node corresponding to the block condition. The data generated within the block is gathered by a block join.

Two conditional nodes marked 'loop' and 'if' in fig.3 correspond to the WHILE and IF statements. C_join in fig.3 receives data from the two branches of IF statement, l_join gets the initial data on his first input and the data produced in the loop on the second one. Join inputs have corresponding control arcs pointing to relevant conditional nodes with Boolean values on those arcs depending on the conditional value for the path the data is coming from.
VHDL process translated to VSG has always a top loop on all the variables whose values are used in subsequent process executions. This assures correct representation of VHDL process simulation behaviour. Representation of the top loop does not differ from the inner loop representation. The initial values of variables defined in the process are fed into the top l_join and the process loop condition is always true. Representation of control information in VSG is better seen in fig.4a where a set of control arcs is shown for nested control blocks of VHDL. This is a very regular representation until no WAIT statements are present in control blocks.

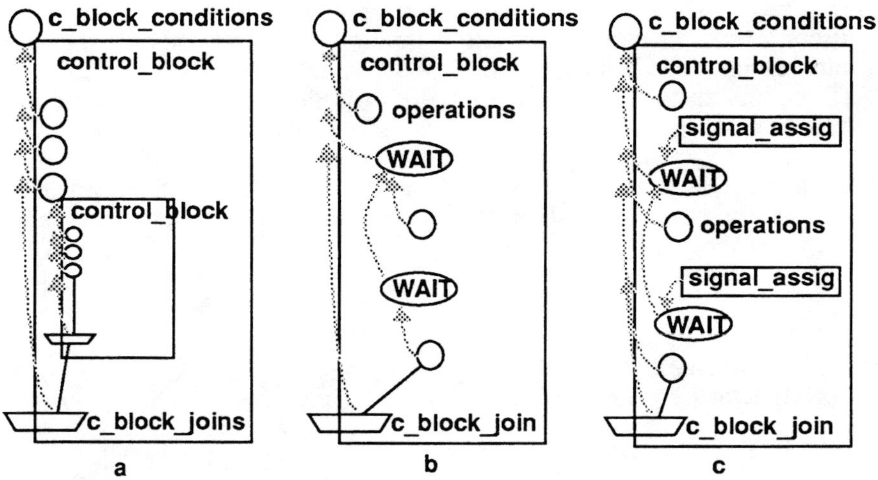

Fig. 4 Representation of control information in VSG

WAIT statement together with signal assignment and reading are VHDL constructs which must be treated with caution. The simulation semantics is such that the simulator executes in zero time all the statements in a process body before it hangs up on the next WAIT. After WAIT condition is fulfilled the following part of the process body is executed and the simulator again stops at the next WAIT. This sequence must be reflected in the representation for synthesis. As shown in fig.4b control arcs of operations in VSG point to proper WAITs and every subsequent WAIT points to its predecessor. The scheduler can retrieve this information and take the imposed sequence into account while scheduling operations. In general, the sequentialisation imposed by WAIT statements can prevent the scheduler from smart reuse of hardware resources for operations which could be moved across WAIT boundaries. A sample of VHDL code where this would be possible is shown below.

```
PROCESS
BEGIN
...
m := m + 1 ;
WAIT on clock UNTIL (clock = '1') AND condition ;
v := a + b + c + d  ;
w : = x + y ;
t := t + z ;
WAIT on clock UNTIL  (clock = '1') AND condition ;
...
END PROCESS ;
```

Some of addition operations in the above code can be moved above the first WAIT where more adders would be available. A general representation of control dependencies as depicted in fig.4c allows optimisation across WAIT boundaries. Applying this principle improves the generated hardware but makes it more difficult for the designer to find correspondence between his specification and the synthesis result. Another problem is that only those operations can be moved which do not depend on input signals.
WAIT statements can have a sensitivity list or if the list is not specified an implicit list is assumed where signals present in the WAIT condition are taken. If the synthesis system is to generate synchronous hardware, all WAITs in the process to be synthesized should have only the clock signal in their sensitivity lists. This simple measure assures that the simulated behaviour of WAITs will match that of the synthesised module.

```
...

W1 : WAIT ON clock
     UNTIL (clock = '1') AND condition1

a := b + S1 ;

S2 <= a ;

W2 : WAIT ON clock
     UNTIL (clock = '1') AND condition2
...
```

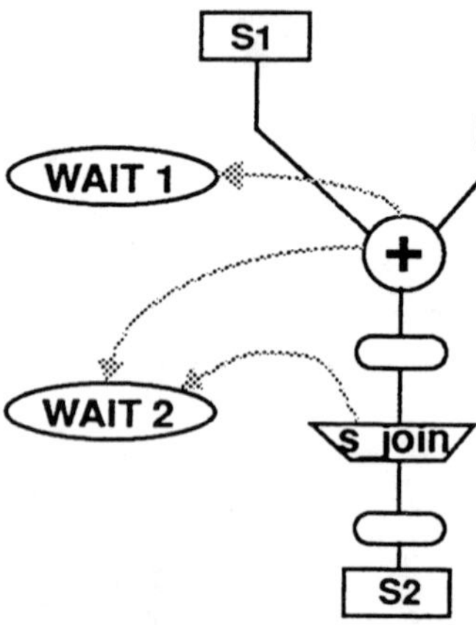

Fig. 5 Representation of signal assignment and sensing

VHDL code and VSG graph in fig.5 explain the representation of signal assignment and sensing. Signal S1 is read from the input port S1 between statements WAIT1 and WAIT2. The operation reading the signal S1 has therefore control arcs pointing to WAIT1 and WAIT2. This operation will be scheduled between those two. Signal S2 is assigned before WAIT2 and is seen on the output port when WAIT2 is executed. Therefore the s_join representing signal assignment has the control dependency arc pointing to WAIT2. This signal assignment will be made in hardware when the synthesised controller

FSM reaches the state corresponding to the scheduled WAIT2.

Inefficiencies of VHDL process description for synthesis

VHDL process was designed by language developers for describing behaviour of hardware without any implementation considerations. It seems that that purpose was fully compatible with what is needed as an input to any high level synthesis system. Unfortunately there are two main problems with VHDL process regarding bahavioural specification for high level synthesis. The first is the strict VHDL timing specification. VHDL timing is absolute without any possibility of setting timing margins or only requirements regarding maximum time needed for execution of a piece of code. VHDL timing is based exclusively on signals, which rules out possibility of specifying timing requirements for operations dependent on variables within a process. Everything between two WAIT statements is executed in zero simulation time. Any hardware implementation will result in some finite execution time. The problem is that it is not possible in VHDL to specify how long execution time is acceptable. This information is needed by synthesis tools as a functional constraint to be obeyed and also by simulation tools if the designer wants to verify functionally the synthesised code versus the behavioural specification by applying the same stimuli to both and comparing signal values. Such verification by simulation is the only practical method available now. The problem there is that the synthesised hardware will generate signals with some finite delay while the behavioural specification gives zero delay. Those two can be perfectly equal functionally but the timing difference makes straightforward automatic comparison of the results impossible.
The second inefficiency is absolute lack of reset possibilities on processes. Resetting in hardware is a very basic operation and should be possible to specify in quite early stages of the design, in behavioural models for synthesis too. Reset is implemented in hardware as an operation almost transparent to other functionality. Reset in a VHDL process can be described by testing the reset signal value in every control block and making exit from loops when reset is active. Testing should be done whenever the process hangs up on WAITs. A sample of VHDL applying this principle is shown below and is ridiculously inefficient to any hardware designer.

```
PROCESS
BEGIN
  ...
  WAIT ..
  IF reset = '0' THEN
    IF func_condition THEN
    ...
    END IF ;
  END IF ;
  WHILE f_condition LOOP
     -- loop body1
     EXIT WHEN reset = '0' ;
     WAIT ...
     -- loop body2
```

```
    EXIT WHEN reset = '0' ;
    WAIT ...
  END LOOP ;
  ...
END PROCESS;
```

Even this elaborate description does not fully cover the functionality which the designer would like to describe. Execution of operations in the loop body before WAIT can be stopped easily in real hardware but not in VHDL process. Something must be done to correct this inefficiency and a proposal for resetting is already on the list of requirements for VHDL 92. Another way to handle resetting of processes stems from observation that resetting is equivalent to restarting the simulator. As restarting the simulator brings the process to its beginning with start values of signals and variables it is functionally equivalent to reset in hardware. The reset signal is not needed to be specified by the designer, it can be generated in the synthesis process. Implementation of reset can be synchronous or asynchronous depending on synthesis strategy. This is the approach taken in the SYNT system.

Synthesis of VHDL statements

A VHDL description for synthesis can consist of many interconnected modules(components) which are to be synthesised or fetched from the library of standard components. Mapping the structural part of such a description to the generated hardware does not cause problems. Using entities and architectures encapsulating only one process gives a direct correspondence between functional and physical partitioning if this is needed. Physical partitioning can be further elaborated by low end tools and in general can result in different partitioning. Still that information can be easily presented with entity and architecture constructs of VHDL. Information about partitioning between chips, assigning signals to physical pins and even placement of modules on a chip and chips on a board can be handled with user defined attributes for synthesis. There is obvious need for some standardisation effort regarding those attributes. Standard attributes for synthesis would contribute to uniformity of VHDL models in various synthesis systems.
Another kind of information about implementation of a purely behavioural model can be gathered in a special synthesis package. It should contain conversion functions for various BIT_VECTOR representations of INTEGER and redefinitions of operators for those representations. The most popular representation for INTEGER is unsigned binary and two complement depending on the declared range. In general, the user can define other representations and the synthesis system must be guided which representation to take. This control can be handled by attributes.
The essential problem of synthesis is converting a VHDL process to a hardware module preserving the simulated behaviour. The weakest point of VHDL in this aspect is its lack of timing specification for synthesis. Present constructs as AFTER time_expression and WAIT FOR time_expression are inadequate and have ambiguous meaning for synthesis. AFTER and WAIT FOR clauses specify strict time for simulation. That can be interpreted as max time specification for synthesis but does not hold in general [5]. If WAIT FOR

is used for time specification it can not be used for process synchronisation. Even with such restrictions usage of AFTER and WAIT FOR is limited by discrepancies between synthesis and simulation interpretation of control sequence in VHDL process. Present high level synthesis systems usually assume a general architecture of the synthesised module consisting of a controller and a data path. The controller is a single finite state machine having only one state active at a time. The single FSM controller is not the best choice for VHDL process synthesis because of its obvious limitation in handling parallel events. VHDL simulation semantics is such that WAIT statements are executed in a simulation process concurrent to the simulation process controlling time. If the synthesis system could preserve this semantics it would be easier to handle timing of signals with AFTER clause. The problem is illustrated below with a simple example.

```
...
S1 <= A*B AFTER 100ns ;
WAIT UNTIL S2 = '1' ;
A := S1 ;
...
```

The synthesised hardware based on single FSM controller will always read a new value of S1, while in simulation an old value can be read if S2 has changed within 100ns. The proposed solution [5] is backannotation of the 100ns delay to the code by inserting WAIT FOR 100ns as shown below.

```
...
S1 <= A*B AFTER 100ns ;
WAIT FOR 100ns ;
WAIT UNTIL S2 = '1' ;
A := S1 ;
...
```

There is no good solution to the problem of timing specification for synthesis in VHDL until a more sophisticated controller models are available and the meaning of AFTER is redefined with synthesis in mind.

As it was stated earlier, most of VHDL sequential constructs can be treated in high level synthesis by standard means. Operators are scheduled and bound to functional units. Joins resulting from VHDL control block statements are implemented as multiplexers. Control sequence and control values for functional units, registers and multiplexers are derived from the scheduled operations and control dependencies of the VSG graph.

There are two VHDL specific constructs which need special treatment: WAIT statements and signal assignments. They are intimately related to each other by virtue of the simulation semantics of VHDL. Representation of these constructs was already discussed.

Lets analyse how much the synthesis result will differ from the simulation model even if that model is made with "synthesis in mind", knowing how the synthesis system handles VHDL timing. Fig. 6 shows a simple example of VHDL code to be simulated and synthesised.

According to the simulation semantics all the code between the two WAIT1 and WAIT2 statements is executed in d==delta time which is infinitely small, signal assignments are scheduled in signal drivers for updating and the process is hanged up. Signals take up new values when the process hangs up but because d is infinitely small this happens virtually at the time of execution of the first WAIT1. How can that behaviour be implemented in real hardware ? For the first, operations take time to execute on hardware units, for the second if the hardware is synchronous (and this is a standard and good industry practice) new signals will be seen earliest on the next clock edge, when the condition on WAIT2 is fulfilled. What can be assumed is that in a synchronous hardware implementation of a process, the d==delta is equal to the clock period. It is only a similarity between those two, not one to one correspondence because execution of any, even very complex code is done in one d for simulation, but can take many clock cycles in real hardware. Such an interpretation of delta allows to get at least some timing correspondence between VHDL process specification and the synthesised synchronous hardware.

VHDL code

```
...
W1 : WAIT ON clock UNTIL (clock = '1') AND cond1 ;
y := a + 1 ; -- can be executed in one clock cycle
S1 <= y ;
W2 : WAIT ON clock UNTIL (clock = '1') ;
... code producing z, executed in 2 clock cycles in hardware implementation
S1 <= z ;
W3 : WAIT ON clock UNTIL (clock = '1') AND cond3;
S1 <= 11 ;
W4 : WAIT ON clock UNTIL (clock = '1') ;
...
```

simulation timing

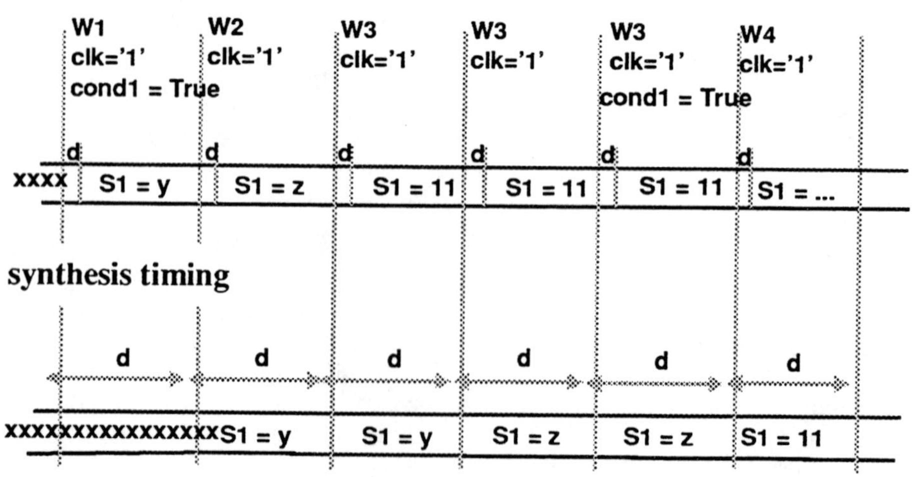

Fig.6 Simulation versus synthesis delta delays

Another problem encountered with VHDL signals is their latching, fig.7 shows a simple example of VHDL code and the corresponding scheduled VSG graph.

```
y := a + 1 ;
... some code using y
S2 <= y ;
y <= a + b ;
...
W1 : WAIT ON clock UNTIL
        (clock = '1') AND condition ;
...
```

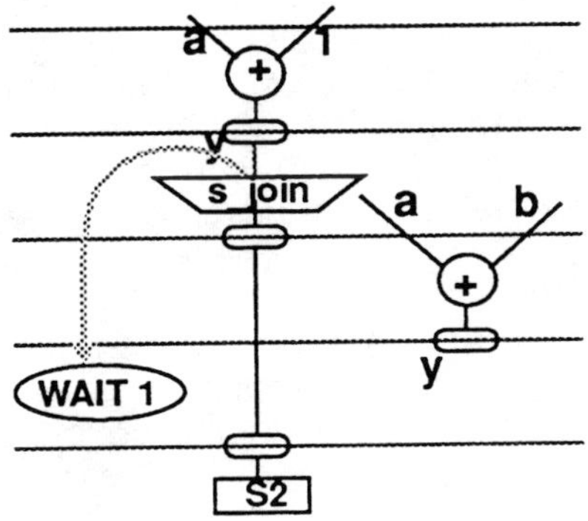

Fig.7 A simple example of latching

Signal S2 must be latched when assignment statement is executed because variable y will later change the value. The latched value of S2 should not be seen on the signal line before the first coming WAIT statement is executed. It means that the old value of signal S2 should be stored in another latch and enabled on the S2 line. We can try to avoid the second latch by allowing the new value of S1 be seen on the line directly when it is assigned. Although diverging from simulation semantics it would not cause any problem in this particular situation. Unfortunately it can not be a general solution, a more complex situation in the VHDL code shown below can not be solved without double latching.

```
PROCESS  BEGIN
   ...
CASE cond1
   WHEN ...
   S1 <= a + 1 ;
   WHEN ...
   S1 <= a - 1 ; --latch here
 ...
    END CASE ;
...
IF cond 2
   WAIT ...  ;      --here the new value of S1 can be seen
ELSIF
    a := a + 10 ;
    S1 <= a ;        --else the value is over-written
END IF ;
   WAIT ...     ;     --and seen here
END PROCESS ;
```

If the new value of S1 was allowed on the signal line already in the upper case statement and cond2 was not fulfilled, the behaviour of the synthesised hard-

ware would differ from the specified one. This leads us to a general double latching principle : signal values must be double-latched in order to conform with simulation semantics of VHDL. There are many situations when the designer does not care about some values, it is quite possible that he would accept the output signal changing its value as in the previous example, if that value was irrelevant to him within some time. Unfortunately that is not possible to specify in VHDL and a safe way to handle synthesis is to trade hardware efficiency for correctness of synthesised hardware. Still there are situations when double latching can be optimised. A general rule can be formulated saying that if the signal assignment is directly followed by a WAIT statement or can be moved to it without changing code functionality the latch in the s-join can be removed. This rule allows to save registers in most of the encountered situations.
Fig.8 explains creation of control sequence from scheduled WAIT statements.

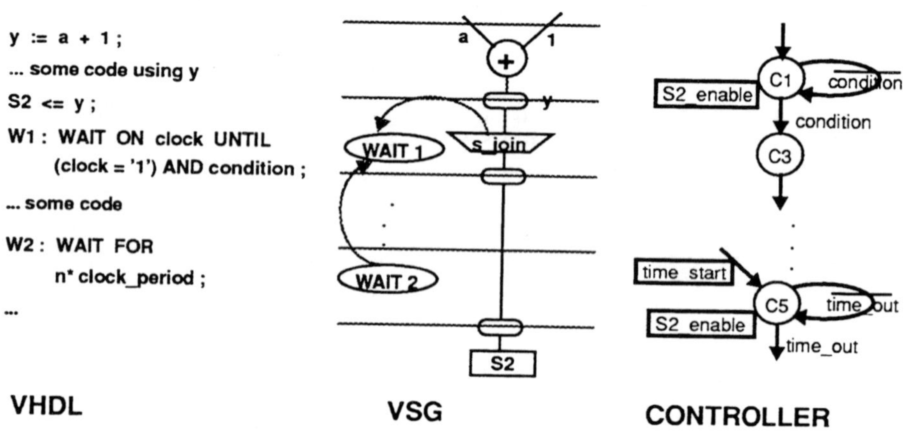

Fig. 8 Extraction of control sequence from scheduled WAIT statements

The scheduled WAIT ON clock UNTIL (clock_edge) AND condition statement is recognised by the controller builder which creates a controller state C1with an internal conditional loop. The condition on the loop corresponds to the WAIT condition. Port registers are enabled in this state putting new values of signals on signal lines. The controller leaves C1 state when the UNTIL condition is fulfilled. The scheduled WAIT FOR n*clock_cycle statement is converted to a FSM state C5 where the controller waits for a signal from a timer. The timer is started by the time_start signal generated by the controller on the transition from the previous state to the C5 state. Since time expression in the WAIT FOR statement is limited for synthesis purposes to n*clock_period - number of clock cycles the timer is also a synchronous device and no delays than the clock cycle are generated. In order to conform to the simulation semantics port registers are enabled in state C5 putting new values of signals on signal lines. The time_out signal from the timer causes the controller to leave this state. A possible optimisation of this solution is to make a trade off between using a separate timer and adding additional states to the controller FSM.

Acknowledgments

Results reported in this paper were brought up together with Mats Fredriksson, Ahmed Hemani, Kurt Nordquist and Lars Ola Stengard, members of the synthesis group at Swedish Institute of Microelectronics. A part of this work was supported by The Swedish Board for Technical Development in the IT4 VHDL CAD Tools project.

References

[1] J.Lis and D.Gajski, Synthesis from VHDL, Proc.Int'l Conf. Computer Design, IEEE Computer Soc.Press, Los Alamitos, Calif.,1988.
[2] M.Crastes, A.Fonkoua, HDLs and Logic Synthesis, Esprit Project 2072 Report.
[3] R.Camposano, L.F. Saunders, R.M.Tabet, VHDL as Input for High-Level Synthesis, IEEE Design & Test of Computers, March 1991.
[4] A.Postula, SYNT synthesis system, IM internal report.
[5] H.Konuk, V.Pemmaraju, Specifying Timing Constraints in VHDL for Synthesis.
[6] A.Hemani, High Level Synthesis of Synchronous Digital Systems using Self Organisation Algorithms for Scheduling and Binding, Ph.D. thesis submitted to Royal Institute of Technology, Stockholm, Sweden, 1991.

ASIC Design Using Silicon 1076

Robert A. Cottrell

LSI Logic Europe plc
Grenville Place, The Ring, Bracknell, Berks., England RG12 1BP
Phone: +44-344-426544 *Fax:* +44-344-481039
Email: rac@lsi-logic.co.uk *or* rac@dcs1.lsil.com

1. Abstract

Silicon 1076 is an integrated design automation (DA) system supporting a top-down design method from VHDL source code to silicon implementation. This paper describes this design method and philosophy, and demonstrates the requirement for an integrated DA system. Silicon 1076 supports design by gradually transforming VHDL code to lower levels of abstraction, using a variety of synthesis tools as appropriate. Silicon 1076 incorporates a number of different synthesis tools for different styles of source code, integrated with a VHDL simulator which operates from the same source code as the synthesis tools. This requires the provision of consistent libraries for simulation and synthesis at all levels of description. A design management system ensures that the correct design flow is followed. Silicon 1076 also incorporates a proprietary CAD system, so as to cover the entire range from VHDL source code to silicon implementation.

J. Mermet (ed.), VHDL for Simulation, Synthesis and Formal Proofs of Hardware, 135–147.

2. Introduction

Since its adoption as an IEEE standard in 1987, there has been increasing interest in VHDL [1]. This has become much more marked recently, with a number of VHDL simulators available which implement 100% of the standard. Many companies are now implementing design strategies with VHDL at the centre.

In parallel with this growing interest in VHDL, there has also been considerable growth in the use of synthesis tools, which transform a design description from one degree of abstraction to a lower level either automatically or semi-automatically. The first synthesis tools were logic synthesis tools for combinational logic. These have now been developed into sophisticated register transfer level (RTL) synthesis tools, which can synthesise sequential logic from RTL descriptions. VHDL is attractive in this context, because it can be used for both the input RTL description and for the output gate level description. Using VHDL means that simulation can be performed both before and after synthesis, and indeed mixed level simulation can also be performed, with code at RTL and gate levels being simulated together with code at higher levels of abstraction. The same simulation testbench can be used in each case.

Despite the advances in this area, RTL synthesis tools are not yet suitable for all kinds of logic. Whereas they can produce high quality results for random logic such as controllers, they are not so good at generating more regular structures such as arithmetic functions and datapaths. Specific synthesis tools, or generators, are available for such functions, but they need to be called explicitly by the designer.

In addition to this, there has been significant research effort recently into high level synthesis [2]. Such synthesis tools work from a description at a higher level than RTL, with the output typically being an RTL description. The input to such tools is described as algorithmic or behavioural code; the key characteristic of such code is that it does not allocate the operations in the source code to specific hardware resources, nor does it schedule them into particular clock cycles. Indeed, behavioural source code must not make reference to the clock or reset signals, as this would violate the essential rules of synchronous design. Rather, some means must be used to identify to the synthesis tool which signals it is to use as clock and reset for the registers it instantiates.

Silicon 1076 capitalises on all these recent developments, combining them together into an integrated design environment supporting top-down design methods.

3. Context of Silicon 1076

Silicon 1076 is intended to be used for the design of Application Specific Integrated Circuits (ASICs) in the context of systems being designed in a top-down fashion using VHDL. It is anticipated that the system being designed will be described at a high level using VHDL and validated by simulation. This high-level description must then be partitioned by the designer into different components of varying types. These could include standard parts such as microprocessors and memories, ASICs, and even software. In the future, there may be design automation tools available to assist in this process, however at present this partitioning must be performed manually. Once a specific ASIC has been identified and described in VHDL, it must be further partitioned into smaller blocks. Once again, in the future this task may be supported by design automation tools, but at present it must be done manually. When the ASIC design has been partitioned into blocks which are at a level understood by one or other of the synthesis tools incorporated in Silicon 1076, the designer is ready to enter the Silicon 1076 design environment.

4. Coding Style and Synthesis Tools

One key aspect of the synthesis strategy used by Silicon 1076 is to direct code to the most appropriate synthesis tool. The synthesis tools supported by Silicon 1076 at present are a general purpose register transfer level (RTL) synthesizer, specialised point synthesis tools for memories and logic functions such as adders and multipliers, and an interactive architectural synthesis tool for high level synthesis. The coding styles for these tools are slightly different.

In all styles of code, it is important that all ports on entities us the same VHDL types, to permit entities and architectures in different styles to be mixed for simulation, and to permit the same testbench to be used throughout. As the LSI Logic technology libraries are described using the T_WLOGIC type [6], all ports on entities must be T_WLOGIC, T_WLOGIC_VECTOR, or a subtype of one of these. It would be perfectly possible for the synthesis tools to take input in terms of BIT and BIT_VECTOR, for example, but it would not then be possible to use the same testbench at all levels. Therefore in the source code, T_WLOGIC and T_WLOGIC_VECTOR must be used, but constants must be described in terms of F0 and F1 only, equivalent to BIT values '0' and '1' respectively, for compatibility with the synthesis tools.

In the future, the technology libraries will migrate to the 9-valued logic system being proposed as an IEEE standard once it has been fully ratified. This includes a specific "don't care" value, which can potentially be used by the synthesis tools to achieve better optimization.

4.1 INTERACTIVE ARCHITECTURAL SYNTHESIS

Silicon 1076 includes an interactive architectural synthesis environment [3] [4], which supports trade-offs between silicon area, clock period and number of clock cycles. Source code for this tool is a behavioural subset of VHDL. Each behavioural entity is separately synthesised to RTL. Behavioural entities are identified by attaching an attribute to the appropriate VHDL architecture. The clock and reset signals to be used by the synthesizer are likewise identified by attributes. These will be used by the synthesis tool when it writes the RTL code.

Behavioural source code can make no reference to the signals used as clock and reset. Making reference to the clock signal would violate the fundamental rule of synchronous design, namely the separation of clock and data signals. The same applies to the reset signal. There are other restrictions which apply to the behavioural code which can be handled in Silicon 1076. An architecture described in the behavioural style may only have one process, and only a limited number of VHDL constructs is supported within this process. There can be no other concurrent statements or processes. In the general definition of behavioural code, there can be separate processes, as long as any signals used for communication between them are only used for synchronisation of the processes, and not for data transfer. Such data transfer would imply structure, and the code would no longer be strictly behavioural. However, at present Silicon 1076 only accepts single-process behavioural code.

The range of VHDL constructs accepted within behavioural code is somewhat limited, but there are plans to ease these restrictions in the future. At present, function and procedure calls are not supported. When these are included in the future, the user will have the option of expanding them rather like a macro, or scheduling and allocating them as a hardware resource. IF and CASE statements are supported, as are WHILE loops. A limited subset of the WAIT statement can be used for synchronisation.

As behavioural code makes no reference to the clock signal, although it must communicate with other entities, there is a need for some means of synchronising these entities. This is most easily achieved by using some form of handshaking. For example, the following behavioural process uses START and BUSY signals for handshaking:

```
process
begin
   BUSY <= F0;
   wait until START = F1;
   BUSY <= F1;
   C := A + B;
   D <= C * E;
   wait until START = F0;
   BUSY <= F0;
end process;
```

An entity which communicates the entity incorporating this process should put the required values on the input signals A, B and E, then set START to F1, wait for BUSY to go to F1, then set START to F0 and wait until BUSY goes to F0, when the result will be available on output signal D.

The architectural synthesis tool will schedule the operations to specific clock cycles and hardware resources, and produce an RTL description of the block.

4.2 REGISTER TRANSFER LEVEL SYNTHESIS

Register transfer level (RTL) code is essentially where logic is defined in terms of combinational logic and latches. However, current RTL synthesis tools do not require latches to be explicitly instantiated: they can be inferred from sequential code. Combinational logic can be described using concurrent assignment statements, but can also be described using a certain style of sequential code [5].

Processes which are intended to result in combinational logic should be sensitive to all their input signals. Additionally, all output signals should be fully specified: they should be assigned each time the process is executed. If a signal is not so assigned, a latch will be inferred. For example, the process:

```
process (a,b)
begin
   if a = f1 then c <= b;
   else c <= f0;
   end if;
end process;
```

will result in combinational logic, and the following process:

```
process (a,b)
begin
   c <= f0;
   if a = f1 then c <= b;
   end if;
end process;
```

will result in precisely the same logic. However, the following process:

```
process (a,b)
begin
   if a = f1 then c <= b;
   end if;
end process;
```

will result in a latch being inferred, as the execution of the process may result in c being left unchanged.

It is also possible to infer edge-triggered flip-flops rather than level-sensitive latches, such as in the following process:

```
process (a,b)
begin
   if a'event and a = f1
   then c <= b;
   end if;
end process;
```

In this case, there can be no activity in the *else* clause of the *if* statement, as this could have no meaning in hardware.

It is also possible to define completely synchronous processes, where the first statement is a wait statement. This must be the only wait statement in the process. In this case, the process will only execute on one edge of the clock, and can therefore be synthesised as synchronous sequential logic. An example of such a process follows:

```
process
begin
   wait until clk = f1;
   a <= b + c;
end process;
```

There are a number of restrictions which apply to the code which can be accepted at the RTL level. For example, processes without a WAIT statement are always synthesized as if they are sensitive to all input signals, and therefore should be written as such. In processes with a WAIT statement, there may be only one such statement, and it must be the first statement in the process. FOR loops are supported, provided that the range is globally static; they are "unrolled" for synthesis and hardware is generated for each iteration of the loop. Other loops, such as WHILE loops, are not supported.

4.3 POINT SYNTHESIS TOOLS

General purpose synthesis tools operating from RTL code as input are generally very good at synthesising controller type functions in random logic. They are not typically so good at synthesising datapaths, and arithmetic logic such as adders and multipliers. While this may change in the future, at present it is advantageous to make use of dedicated synthesis tools for such blocks. The same applies to instantiating memories. One key aspect of Silicon 1076 is that the same source code can be used for simulation and synthesis, so it is necessary to provide access to simulation models of the functions to be generated, and to the generation tools themselves, from the same source code.

This is achieved in Silicon 1076 by using concurrent procedure calls. These are supported by simulation models, and are also understood by the synthesis manager which provides the link to the generators themselves. The following code is all that is required to synthesis a carry look-ahead adder:

```
ADDER_CLA ( A=>A, B=>B, CI=>C0,
    CO=>C2, S=>HASH_CAPACITY);
```

A, B and HASH_CAPACITY are vectors whose length determines the bit width of the synthesised adder, and C0 and C2 are single bits. When simulating this code, a call will be made to a library procedure. During synthesis, the synthesis tool will be called and a structural VHDL description, in terms of cells from the technology library, will be produced.

5. Design Flow

Silicon 1076 supports a top-down design method, and hence the design flow is also top-down, although Silicon 1076 begins part way down the flow of system design. The input to Silicon 1076 is a VHDL description of a single ASIC, where all the code is in a style understood by one of the synthesis tools.

The VHDL description of the ASIC needs to be hierarchical. The top level of hierarchy will include direct instantiation of input/output buffers from the appropriate technology libraries, and instantiation of blocks which represent the top-level partitioning of the design. It is also possible to include RTL code at this level, although it is probably better practice to include this at a lower level of hierarchy. The description will then continue to lower levels of hierarchy, until each block is defined in terms of behavioural code, RTL code, calls to point synthesis tools, or instantiations of cells from the technology library. All forms of code may be mixed in any one VHDL architecture, with the exception of behavioural code which must be in a separate VHDL architecture from all other styles of code.

5.1 REGISTRATION

The files containing the VHDL source code are registered with the system. It works out any dependencies between files, and ensures that each source file is compiled whenever necessary, while avoiding unnecessary compilations. It also keeps track of which architecture is current for a particular entity, as architectures at one level are transformed to lower levels by synthesis.

5.2 CODE ANALYSER

The source code is first run through a code analyser. This verifies that the code is syntactically correct VHDL, and also that it is can the synthesized by one or other of the synthesis tools. The user is not forbidden from proceeding with code which cannot be synthesized, but it is useful to be warned early on of likely future problems.

5.3 HIGH LEVEL SIMULATION

The first stage in the design flow is high level simulation: the simulation of the user's original source code. The purpose of this simulation is to validate the description. If the user is defining the whole system using VHDL, then this description can be simulated in the context of that system. The user may also have a higher level description of the ASIC, but which cannot be directly synthesised, in which case the aim is to verify the manual design of the code for synthesis. In any case, it should be the aim of the user to develop a testbench which will be able to validate the design at this level, and also verify the synthesis processes by being used later on to simulate the results of these processes.

5.4 INTERACTIVE ARCHITECTURAL SYNTHESIS

Having validated the VHDL source code for a design, the next stage is to transform any behavioural style code into RTL. This is done in Silicon 1076 by interactive architectural synthesis, which is performed separately on each behavioural architecture. The first stage of this process is for the code to be analysed and converted into the form of a complex state machine: a controller and a datapath. This description is shown graphically to the user in the form of a control flow graph and a number of data flow graphs. An example of both is given in Figure 1. The control flow graph represents the essential control flow of the source code, with each box representing a separate state. There is a data flow graph for each state box, in which the square boxes at the top represent data inputs in that state, the square boxes at the bottom represent data outputs. The circles represent atomic operations in the source code, which need to be scheduled and allocated. The arrows on the data flow diagram represent data dependencies.

Having performed the initial analysis, the synthesis tool uses a process of clustering and partitioning to develop potential hardware architectures for investigation. Atomic operations are hierarchically clustered, using knowledge embedded in the system to determine which operations may make sense to combine onto a single piece of hardware. The clustering continues until all operations are contained within a single cluster. Each level in the clustering hierarchy represents a potential hardware architecture for implementing the source code. That architecture consists of the minimum hardware resource required for each cluster. In this way, a reasonable small number of architectures is suggested, from the most parallel (specific hardware for every atomic operation) to the most serial (the minimum hardware required). This is illustrated in Figure 2, which shows a fragment of VHDL source code, a cluster tree for that code, and a graphical representation of three possible hardware architectures.

The synthesis tool then proceeds to estimate the silicon area for each potential architecture. This uses a proprietary estimation tool, based on experience of thousands of ASIC designs. In order to so this, a preliminary scheduling and allocation is performed to estimate the required number of registers and multiplexers, but this may be different in the final implementation. Even at this high level, close integration with the final silicon implementation becomes important. It is difficult, if not impossible, to make sensible decisions even at this level without feedback from reasonable accurate timing and area estimates.

The user is then presented with a graphical representation of the relative sizes of the potential architectures. He then selects one such architecture to explore further, by specifying the clock period and running the architecture synthesis program. This performs the tasks of scheduling, allocation and connectivity binding, and produces the RTL code. The resulting architecture is presented back to the user in the form of a scheduled flow graph. The control flow graph never changes, but the rows on the data flow graph now represent sequential clock cycles. Horizontal arrows represent operations which have been chained into one clock cycle; circled occupying more than one row have been scheduled to take more that one cycle.

The number of cycles taken for any state block can be increased above the minimum of one either because of hardware resource limitation, or because the clock period is too short to accommodate all the operations for that state. The user can continue to explore different architectures with different clock cycles until an acceptable compromise is found.

The execution times for the architectural synthesis operations is relatively short, making it a practical proposition to try out several possibilities before deciding on one. This is in contrast to logic synthesis and optimization tools, which typically have long run-times.

The RTL code which results from architectural synthesis makes use of the point synthesis tools for suitable operations.

5.5 RTL SIMULATION

Architectural synthesis is followed by simulation at the register transfer level. Exactly the same testbench is used as for high-level simulation, but with a different configuration.

It is not always easy to determine the equivalence of the results of RTL simulation and high-level simulation, because operations may be scheduled in different clock cycles. It is crucial that the testbench is well designed to ensure the correct operation of the system. The aim must be to verify functional equivalence in spite of timing differences.

5.6 RTL SYNTHESIS

The process of RTL synthesis is the transformation of an RTL description to a netlist in terms of elements from the appropriate technology library. In the case of Silicon 1076, this process is controlled by a synthesis manager, which directs the code to the appropriate synthesis tool. This may be one of the special purpose point synthesis tools, or the general purpose RTL synthesizer. This is essentially an automatic process with no user interaction.

5.7 LOGIC OPTIMIZATION

The result of the RTL synthesis process could arguably be implemented directly, but it is almost certain that it can benefit from optimization. The user needs to specify appropriate constraints to be applied in the optimization process, which can be in terms of area, delay, clock period, maximum rise/fall times, etc. The process of logic optimization can take considerable resources in terms of computer time and memory; it is worth developing a good set of constraints and then leave the computer to get on with it.

5.8 GATE LEVEL VERIFICATION

Gate level verification of the design can take place in VHDL, if the estimated gate delays are back-annotated into the VHDL code. However, most VHDL simulators are not well suited to gate-level simulation, and it is easier to use a conventional logic simulator. Both are supported in Silicon 1076; VHDL gate-level simulation is probably only really useful as a means of achieving mixed-level simulation.

6. Conclusions

A top-down method for ASIC design based on VHDL has been presented. The essential aspects of this design method are the use of a variety of synthesis tools operating at different levels, with all the synthesis tools closely linked to simulation so that precisely the same source code can be used for simulation and synthesis. This requires the provision of compatible libraries for simulation and synthesis. Whereas the method could be applied by using a variety of tools independently, there is a very great advantage in having all the tools and libraries integrated into a consistent, well controlled environment such as Silicon 1076.

In the future, synthesis tools will become more powerful. However, the concept of using different synthesis tools for different purposes is likely to be around for some time. The number of different synthesis tools incorporated into an environment such as Silicon 1076 is likely to increase rather than decrease, especially as synthesis becomes useable at ever higher levels in the design process.

It is anticipated that some of the problems at present, such as the limitations on the styles of code which can be accepted will be addressed by developments in the synthesis tools themselves. Other problems, such as the adoption of standard types for describing logic, will be addressed by the various standardisation bodies.

7. References

[1] IEEE Std 1076-1987, "IEEE Standard VHDL Language Reference Manual", IEEE, New York, 1988.

[2] M. C. McFarland, A. C. Parker and R. Camposano, "Tutorial on High Level Synthesis", Proceedings of the 25th Design Automation Conference, pp 330-336, Anaheim, CA, June 12-15, 1988.

[3] Vijay Nagasami and Carlos Dangelo, "Explorer: A Trade-off Analysis Tool for Architecture Synthesis", Proceedings of the 5th annual European Computer Conference, Bologna, Italy, May 13-16, 1991, pp327-330.

[4] Neerav Berry, "A flexible architectural synthesis paradigm for VHDL", VHDL Users' Group Spring 1991 Conference, pp47-53, April 8-10, 1991, Cincinnati, Ohio.

[5] "Introduction to HDL-Based Design Using VHDL", Synopsys, Inc., 1990.

[6] David R. Coelho, "The VHDL Handbook", Kluwer Academic Publishers, 1989.

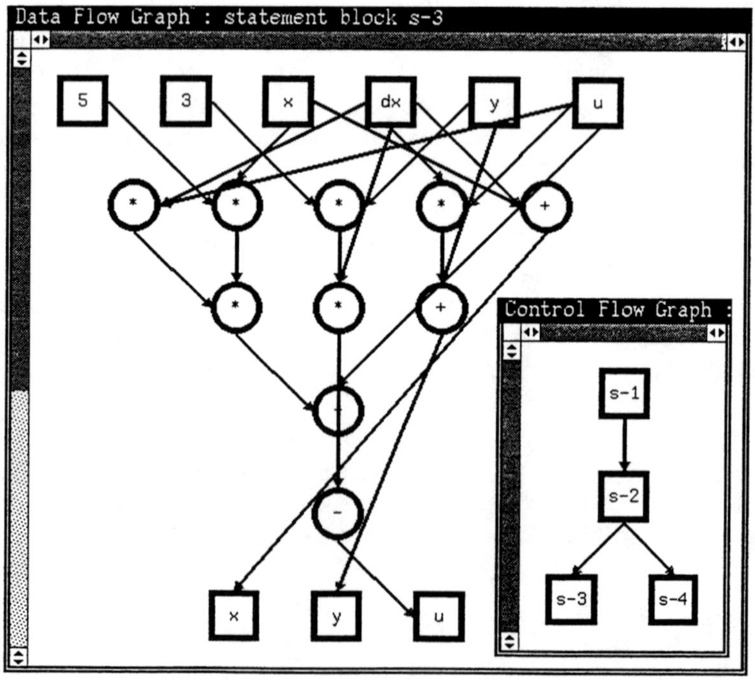

Figure 1 *Control and Data Flow Graphs*

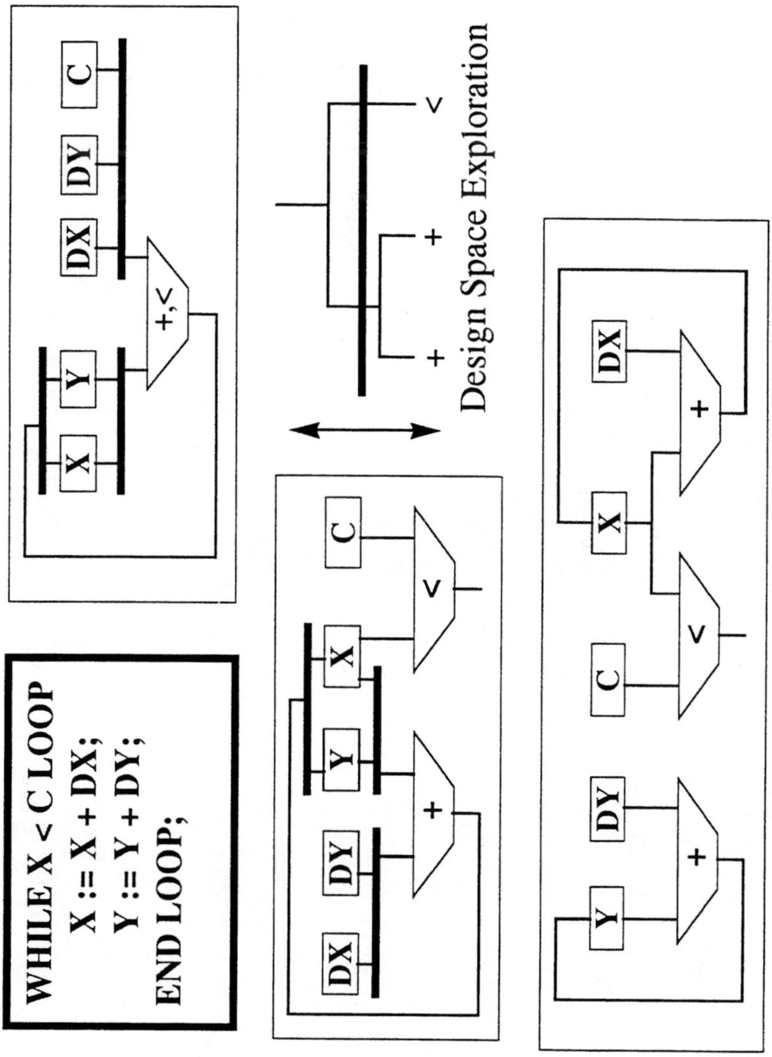

Figure 2 *Example of Clustering and Partitioning*

Generating VHDL for Simulation and Synthesis from a High-Level DSP Design Tool.

LARS LUNDBERG

Comdisco Systems, Inc., 919 East Hillsdale Blvd., Foster City, CA 94404, USA
Phone: +1 (415) 358-3658, Fax: +1 (415) 358-3601, Internet: LL@CSI.com

1. Abstract

This paper describes how a commercial tool for design of digital signal processing systems at the algorithmic level has been adapted to generate VHDL that can serve as input for simulation and synthesis. Special attention will be paid to how the VHDL has to be generated in order to allow efficient synthesis using a popular commercial tool. At the same time the VHDL must be flexible so that it can provide input to other synthesis tools in the future, and generic enough to simulate on any commercial VHDL-simulator.

After a brief introduction to the background and purpose of the project, a simple design example is given as proof of concept. The focus of this paper is on how VHDL is generated from the front-end DSP design tool. The following sections offer a closer look at how this VHDL is used as an interface to synthesis and simulation. A short discussion of the current limitations of the tools and possible future directions serves as the finishing point.

2. Introduction

Recent advances in hardware synthesis have made it attractive to connect from system level design tools directly to tools for hardware implementation. A catalyst in this process has been the hardware description language VHDL.

The joint Comdisco Systems/Synopsys tool mentioned in this paper provides a top-down design path for DSP systems that can dramatically shorten the time from specification to implementation. VHDL plays the crucial role of an interface between the tools from the two companies.

J. Mermet (ed.), VHDL for Simulation, Synthesis and Formal Proofs of Hardware, 149–161.

The concept of using output from a high level design tool as input to a hardware synthesis tool will become increasingly popular as tools capable of synthesis from behavioral descriptions become available. Tools that were used for design and verification of behavior at the system level can now serve as front-ends to tools leading all the way down to hardware implementation. Being an open standard, VHDL is a good choice for interfacing system level and synthesis tools.

Textual VHDL is gaining acceptance as a general purpose front-end to the design process. However, in many cases there are more powerful ways of entering a design. Use of VHDL as a back-end to a more attractive design entry tool isolates the designer form textual VHDL and at the same time maintains the benefits of VHDL for simulation, synthesis and documentation.

Digital Signal Processing (DSP) is a field of applications well-suited for automated top-down design. The reason for this being that the design entry signal-flow block-diagram is a well established front-end to the DSP design process. Typical design elements such as adders and multipliers are regular and therefore easy to generate automatically from parameterized descriptions. Generation of VHDL from a signal-flow description of a system is relatively straight-forward.

3. A Top-down DSP Design Example

This example is an introduction to the concept of top-down DSP design. It provides a general overview of the design process before going into detail in the next few sections. The example is very simple, and is meant to serve more as a proof of concept than anything else.

A notch filter is a very narrow band-stop filter. Implemented as a simple second order section ("biquad") it can efficiently filter out a specific frequency component in a signal. Designing a digital notch filter to be implemented in fixed precision hardware can prove a little trickier. Overflow and quantization due to limited width signal paths and coefficients can significantly change the characteristics of the filter.

The Hardware Design System (HDS) extension to the Signal Processing Worksystem (SPW) from Comdisco Systems allows for experimentation with fixedpoint effects at a system level, before going to implementation. After simulation and optimization it was found that 10-bit data-paths and 3-bit coefficients would give an attenuation of more than 53 dB at the notch frequency.

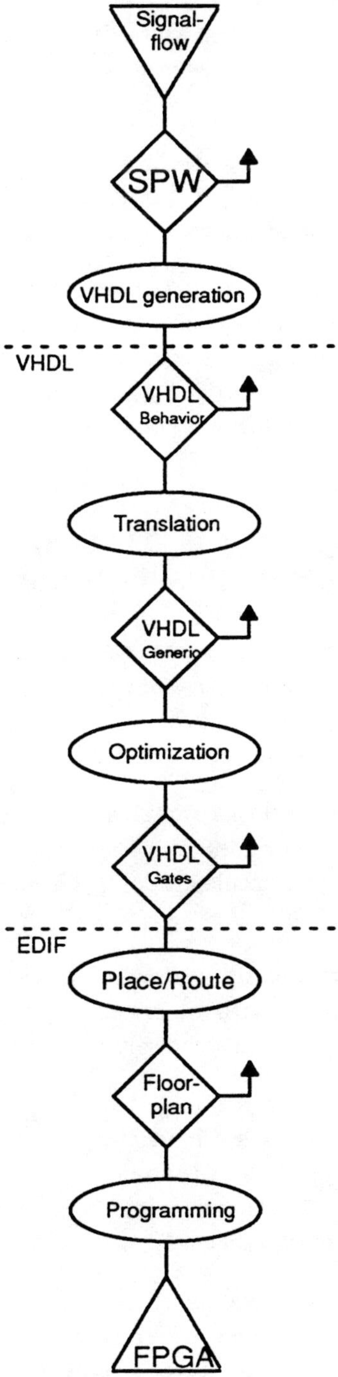

Fig. 1 The top-down flow in the design example.

The next step in the design process was to generate a bit-accurate behavioral VHDL description of the notch biquad. This behavioral description was read into the Synopsys synthesis tools and a decision was made to target an Actel field-programmable gate-array (FPGA).

The Synopsys tools communicates with the Actel tools through an EDIF netlist. After placing and routing the design, an FPGA programming device was used to program the chip. The chip function was verified on a breadboard test bench.

The entire design flow can be found in Fig. 1 on the previous page. Throughout the design process the design was simulated or analyzed at different levels (illustrated with diamonds in the figure) using the same test-vectors that were used to validate the design at the top level. All tools used in the process ran on a single workstation, including the FPGA-programmer that was connected to the workstation through a SCSI port.

4. The DSP Design Tool

SPW is a graphical, interactive simulation tool for the design, analysis and testing of signal processing systems. In the SPW a designer enters blocks and connects them with wires to describe a system much the same way as in a schematic capture tool. The level of abstraction is higher in SPW where typical design blocks range from adders and multipliers to FFT-blocks, and the wires correspond to signal-flow arcs representing anything from buses to vectors of floating point numbers.

There are more than 300 building blocks from areas such as communications, radar and image processing available to the designer. The system also allows the designer to define new blocks, either by encapsulating design hierarchy or by describing the function of the block in C or FORTRAN.

Representation of DSP systems as signal-flow block diagrams is a widely used method. The signal-flow paradigm can be described as synchronous data-flow. This allows for static scheduling in the simulator resulting in quick execution of relatively complex systems.

The signal flow block diagram of the notch filter from the design example is shown in Fig. 2 below. The string '<10,0,t>' found on the output of all blocks means that the signal net is 10 bits wide and is to be interpreted as having 0 integer bits and the representation is two's-complement, ie. a ten-bit two's-complement fractional.

Another important feature of the SPW is its signal editing capabilities. Through the Signal Display Editor (SDE) the designer can create complex signals to be used as test-vectors for his design. The SDE can also be used to view and analyze the results of a simulation.

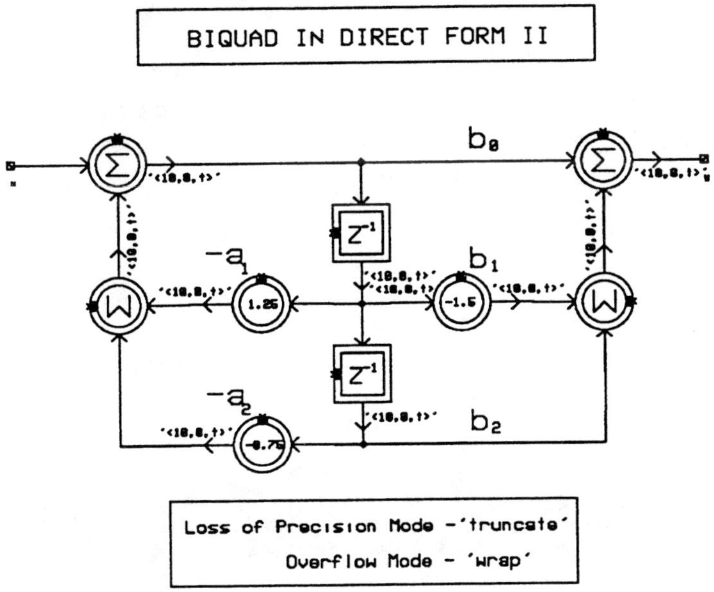

Fig. 2 The signal-flow block diagram of the filter.

In Fig. 3, at the bottom of the page, two signals appear as displayed in the SDE. The upper signal is a couple of milliseconds worth of sampled speech that has been distorted by adding a sine wave to it. The bottom signal is the result of feeding the top signal through the notch filter in the example — the distorting sine wave has been filtered out.

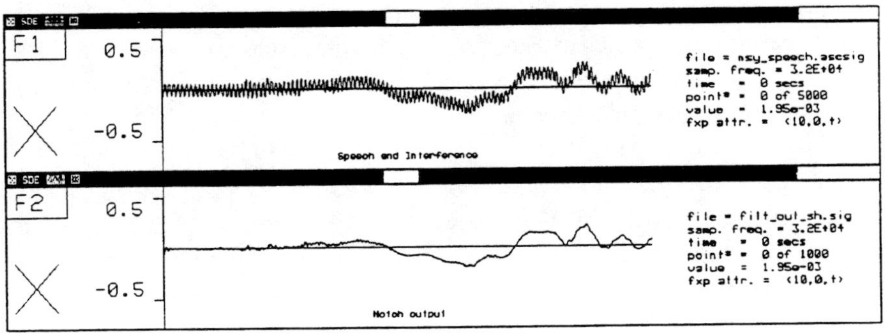

Fig. 3 Speech mixed with a sine wave (F1) and the result after running F1 through the notch filter (F2).

The test bench used to test the notch filter is displayed in Fig. 4 below. The source block reads a signal file from disk, and the sink block writes the result from simulation to disk. Note that the number of signal bits and the number of coefficient bits appear on the filter symbol. Modifying these values on the top-level symbol will change the corresponding values throughout the design, thus simplifying experimentation with fixed-precision effects.

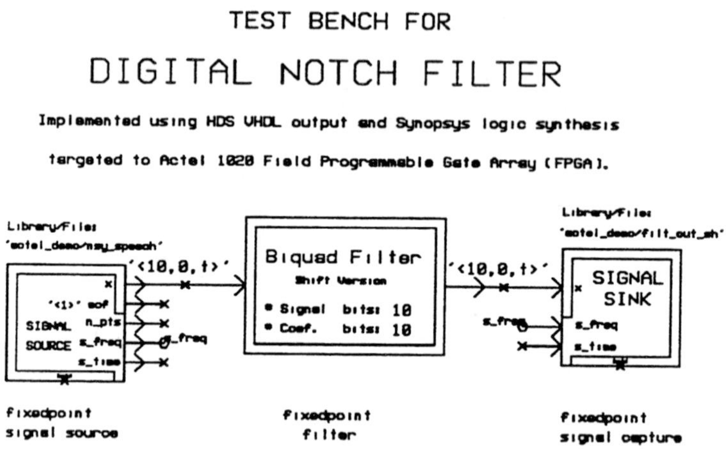

Fig. 4 The biquad notch filter in its test bench.

5. VHDL Generation

There were two goals set up for the VHDL generation tool:

First, the generated VHDL should, when *simulated* in any commercial VHDL simulator, give results that could easily be compared to the results produced by the SPW simulator. This is invaluable for cross-checking the correctness of generated VHDL against system-level block-diagrams.

Second, the VHDL generation was viewed as a way to simplify implementation from systems described in SPW. It was, in other words, important that the VHDL generated was both *synthesizable* and easily understood by those choosing a manual implementation path.

How these two goals were met will be discussed in the following two sections. There were also *implementation issues* that affected the way the VHDL generation was designed.

In the first release of the VHDL generation tool, only blocks capable of dealing with fixed precision signals will be supported. These blocks are found in the HDS block library. The heart of the HDS block library is a subroutine library for fixedpoint arithmetics. This fixedpoint library has been translated to VHDL which means that the VHDL block models appear similar to the existing models in C/FORTRAN. This also means that expansion of the HDS block library can be done without compromising the VHDL support.

The VHDL fixedpoint arithmetic package is an important result from this project. It differs from other arithmetic packages defined on Bit_Vectors (such as the synopsys package from Synopsys) in that it has a notion of a binary point and deals with overflow and loss-of-precision. These features are crucial for correct modelling of fixed-precision hardware.

The basic design unit in SPW is a block. Thus, it was natural to encapsulate the behavioral VHDL description of a block in a VHDL procedure call. The procedure declarations for all supported blocks in the HDS library reside in a separate package.

The behavioral VHDL description of the notch filter in the design example is shown in Fig. 5 on this page. The variables declared as "TC(0 downto –9)" are 10-bit fractional two's-complement numbers. Negative indices represent fractional bits. The other supported datatype today is UN, which is an unsigned fixedpoint number. The sample clock, which is implicit in the signal-flow block diagram, is introduced as a port in the entity.

It is also possible to generate a structural VHDL description from systems designed with HDS blocks. The component declarations are found in a third package along with their corresponding entities and architecture bodies. The structural VHDL description of the test bench used for testing the notch filter is found in Fig. 6 below. Note the reference to the clock and to the notch behavior through component instantiations.

The package structure described above can be visualized as in Fig. 7 below. The left leg is part of the HDS 1.5 product and allows simulation of SPW/HDS systems using any commercial VHDL simulator. The right leg is what Synopsys has added and will be discussed in more detail later.

This layered package structure provides flexibility. Synthesis vendors and other companies using the VHDL generated from SPW/HDS can choose to interface at the level best suited to their purposes. One drawback with the layered structure is that overloading with respect to TC and UN must be provided at all levels. This means a lot more code to maintain. The notion of a "supertype" in VHDL would greatly reduce this problem.

```
-----------------------------------------------------------
--
-- Behavioral VHDL description of actel_demo/notch_sh
-- (C) Copyright 1991 Comdisco Systems, All Rights Reserved
--
-----------------------------------------------------------

use work.fxp.all;
use work.hds.all;

entity notch_sh is
  port(X_PORT : in TC(0 downto -9);
       sample_clock: in Bit;
       Y_PORT : out TC(0 downto -9));
end notch_sh;

architecture hds_body of notch_sh is

begin
  notch_sh: process

  variable Y : TC(0 downto -9);
  variable X : TC(0 downto -9);
  variable net_2 : TC(0 downto -9);
  variable net_3 : TC(0 downto -9);
  variable net_4 : TC(0 downto -9);
  variable net_7 : TC(0 downto -9);
  variable net_8 : TC(0 downto -9);
  variable net_9 : TC(0 downto -9);
  variable net_10 : TC(0 downto -9);
  variable net_14 : TC(0 downto -9);
  variable net_19 : TC(0 downto -9);
  variable net_20 : TC(0 downto -9);
  variable net_22 : TC(0 downto -9);
  variable net_23 : TC(0 downto -9);

  begin
    wait until (sample_clock'event and sample_clock = '1');

    X := X_PORT;

    delay (net_10, net_9);
    shift (net_9, -2, net_20);
    sub (WRAP or TRUNCATE, net_20, net_9, net_3);
    delay (net_14, net_10);
    shift (net_10, -2, net_19);
    add (WRAP or TRUNCATE, net_19, net_10, net_2);
    add (WRAP or TRUNCATE, net_3, net_2, net_4);
    add (WRAP or TRUNCATE, X, net_4, net_14);
    shift (net_10, -1, net_23);
    add (WRAP or TRUNCATE, net_23, net_10, net_22);
    neg (WRAP or TRUNCATE, net_22, net_8);
    add (WRAP or TRUNCATE, net_9, net_8, net_7);
    add (WRAP or TRUNCATE, net_14, net_7, Y);

    Y_PORT <= Y;

  end process notch_sh;

end hds_body;
```

Fig. 5 Behavioral VHDL description of the notch filter.

```
------------------------------------------------------------
--
-- Structural VHDL description of actel_demo/notch_sh_tb
-- (C) Copyright 1991 Comdisco Systems, All Rights Reserved
--
------------------------------------------------------------

library std, work;
use std.textio.all;
use work.fxp.all;
use work.hds.all;
use work.hds_component.all;

entity notch_sh_tb is
end notch_sh_tb;

architecture hds_body of notch_sh_tb is

  signal sample_clock : BIT;
  signal S_FREQ : INTEGER;
  signal net_3 : TC(0 downto -9);
  signal net_4 : TC(0 downto -9);
  signal net_5 : INTEGER;
  signal net_6 : UN(0 downto 0);

 component actel_demo_notch_sh
  port(X_PORT : in TC(0 downto -9);
       s_clk: in Bit;
       Y_PORT : out TC(0 downto -9));
  end component;

begin
  CLOCK: sysclock port map (sample_clock);

  SOURCE_440: hds_lib_source_t
    generic map ("actel_demo/nsy_speech.ascsig", 10, 0)
    port map (sample_clock, net_3, net_6, S_FREQ, net_5);

  NOTCH_SH_1492: actel_demo_notch_sh
    port map (net_3, sample_clock, net_4);

  SINK_484: hds_lib_sink_t
    generic map ("actel_demo/filt_out_sh.ascsig", 10, 0)
    port map (sample_clock, net_4, S_FREQ, net_5);

end hds_body;

configuration actel_demo_notch_sh_tb of notch_sh_tb is
  for hds_body
    for CLOCK: sysclock
      use entity work.sysclock(hds_body);
    end for;
    for SOURCE_440: hds_lib_source_t
      use entity work.hds_lib_source_t(hds_body);
    end for;
    for NOTCH_SH_1492: actel_demo_notch_sh
      use entity work.actel_demo_notch_sh(hds_body);
    end for;
    for SINK_484: hds_lib_sink_t
      use entity work.hds_lib_sink_(hds_body);
    end for;
  end for;
end actel_demo_notch_sh_tb;
```

Fig. 6 Structural description of the notch test bench.

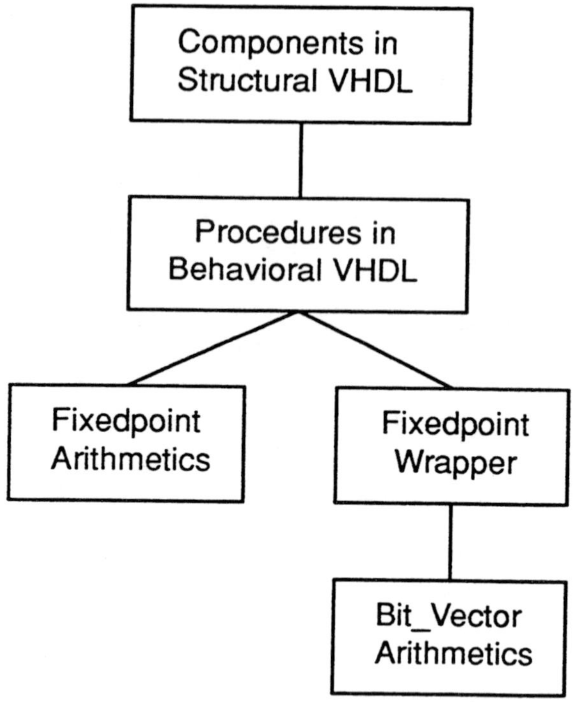

Fig. 7 The current VHDL package structure.

Another problem with VHDL in this context is that unconstrained arrays of unconstrained datatypes are not supported. This means that procedures describing blocks dealing with fixedpoint (TC and UN) vectors are hard to implement. This is a significant restriction that should be addressed in future revisions of VHDL.

6. VHDL Simulation

Simulation is the way to ensure that the VHDL generated still describes the function that was originally intended and validated in the SPW simulation. In an ideal, top-down, world there would be no need for simulation. But for real world designs there will always be reasons for concurrent verification throughout the design process. Even if simple systems like the notch filter could be implemented with validation only at the top level, the designer will often want to verify the design after each step.

One important aspect of this concurrent verification of the design is the ability to re-use the test-vectors that were used to validate the function at the top-level. It is also important that the signals can be viewed differently at different levels of the design. At the top level the magnitude of the signal, represented by a floating point value, is viewed and analyzed. At the gate-level, the binary representation of the signal makes more sense. In the SPW/HDS tool the fixedpoint type allows easy interpretation of signals with real world values. SDE provides the ability to view and analyze the signals the way the user desires.

This functionality is maintained in the VHDL world. By developing VHDL models for the fixedpoint source and sink blocks, the same test bench as on the top level can be used for VHDL simulation. These two VHDL blocks read and write the SPW/HDS signal format. One difficulty when implementing the two blocks in VHDL was that the VHDL TEXTIO package currently has very limited functionality.

The support for VHDL simulation is seamlessly integrated into the SPW/HDS front-end. Provided that a commercial VHDL analyzer and simulator is installed on the system, VHDL simulations can be executed just as easily as the SPW simulations — by the click of a button. The results from the VHDL simulation can be read by the SDE, analyzed and compared to validated test-vectors.

7. VHDL Synthesis

The second idea behind generating VHDL was to enable the use of SPW designs as starting points for automated hardware generation. The generated VHDL can of course be said to provide this ability without any specific modifications — *any* VHDL can be synthesized in principle. Because of the complexity of the synthesis problem it is important to generate VHDL that will have a chance of resulting in usable hardware once synthesized.

To meet this last criteria, Comdisco Systems and Synopsys have jointly developed an interface between the tools from the two companies. For Synopsys this involved adding a fixedpoint package on top of their "synopsys" Bit_Vector arithmetic package. The fixedpoint wrapper adds the notion of a binary point as well as overflow and loss-of-precision handling to the synopsys package. This package structure is depicted in the right leg of Fig. 7 on the previous page.

The interface to the Synopsys tools was used to synthesize the notch filter in the design example and the resulting gate-level design is found in Fig. 8 below. An EDIF version of this netlist provided the input to the Actel FPGA programming tools.

One powerful capability worth emphasizing is the quick re-design times offered by this approach. Changing the width of the signal path from 10 bits to 12 just requires changing a single parameter on the symbol for the notch filter (Fig. 4). Running through the entire design process from top to bottom, *including* verification at SPW-, VHDL behavior- and gate-levels, is done in an afternoon. Targeting another

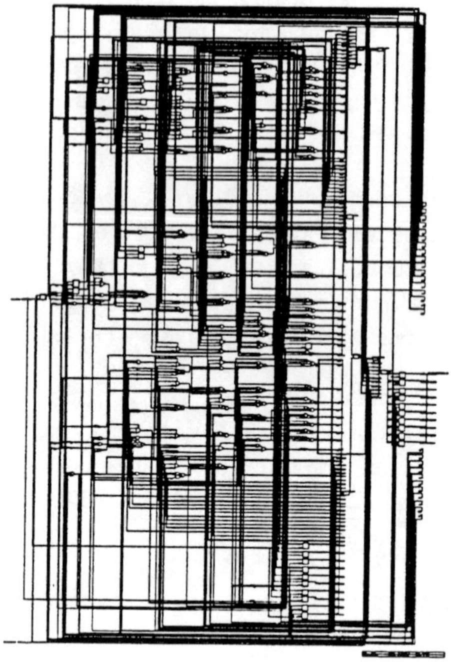

Fig. 8 The result from synthesis of the notch filter.

chip is of course also straight-forward thanks to the Synopsys tools. This capability makes the joint Comdisco Systems/Synopsys tools attractive for rapid prototyping, especially when targeting programmable logic such as FPGAs.

8. Limitations

In the first version of the tool, VHDL can only be generated from systems designed using blocks from the HDS fixedpoint library. These blocks include arithmetic, logic and memory functions that can be used for design of complex fixed-precision DSP systems.

Today the support for high-level synthesis in commercial tools, such as those from Synopsys, is limited. For example, by setting the system clock equal to the sample clock, a fully parallel implementation of the SPW/HDS design is implied. Resource sharing can be introduced by some fairly simple manual editing of the VHDL code before reading it in to the synthesis tool. Ways of doing this automatically will be provided in future versions of the tool.

Another problem, common to DSP hardware design in general, is the fact that some of the most commonly used design elements are very expensive to implement in hardware. A parallel 16x16 multiplier will devour all resources in the latest generation FPGA chips. As silicon gets cheaper and synthesis tools get smarter this problem will become less of an obstacle.

9. Conclusions

Tools for top-down design are fairly new to the market. The reasons for generating VHDL from the SPW DSP design tool is to provide an interface to this new generation of design tools. A couple of trends in future hardware design have been brought into light during the course of this project.

Tools that were previously used by system designers for design and validation of system level behavior will become tomorrow's front-ends to top-down hardware design processes. The catalyst in this process is VHDL, a language with some significant shortcomings today, but still more benefits — the most important being that it is a standard.

There will always be high-volume designs demanding skilled designers and lengthy design processes. But as the complexity of the design increases, even these designers will have to raise their level of abstraction and rely more and more on top-down design. The Comdisco Systems/Synopsys tool relies on continuing progress in high-level synthesis to become successful in production quality DSP hardware design.

10. Literature

An introduction to the use of signal-flow block diagrams and a discussion on limited-precision effects in DSP design can be found in "Digital Signal Processing" by Alan V. Oppenheim and Ronald W. Schafer.

The book "Introduction to HDL-Based Design Using VHDL" by Steve Carlson provides exactly what the title hints. Examples in the book are prepared using the Synopsys synthesis tools and do not necessarily apply to other tools.

Aspects of Optimization and Accuracy for VHDL Synthesis

John Elliott and Paul Harper

Mentor Graphics Corporation
Synthesis, Simulation and Test Division
8005 S.W. Boeckman Road
Wilsonville, Oregon, USA 97070

Introduction

In this paper we discuss some aspects of the synthesis process in our VHDL synthesis tool, AutoLogic VHDL (ALV). ALV synthesizes a subset of the VHDL language, producing a technology independent, gate level design that implements the behavior of the VHDL description.

A fundamental tenet of our synthesis process is to produce efficient designs while maintaining functional equivalence (on the boundaries) between the simulations of the VHDL description and the synthesized result.

In the first half of the paper we discuss some of the subtleties of VHDL simulation semantics and how we treat these issues in the synthesis process. We focus on the implication of clocked storage elements from the VHDL description. In the second half of the paper we discuss our approach to synthesizing efficient designs which target the design goals of the user. In this section we discuss the variety of optimizations employed in the synthesis process whose overall goal is the reduction of functional hardware in the design. We also explain the micro-architectural selection process. This is the part of the synthesis process that generates the architectures of operators to meet overall design goals.

J. Mermet (ed.), VHDL for Simulation, Synthesis and Formal Proofs of Hardware, 163–175.

Ensuring Simulation Equivalence of Synthesis Results

The goal of a VHDL synthesis tool is to produce a gate level implementation from a VHDL description. It is a requirement of synthesis tools that the output of a VHDL simulation match the output of a simulation of the synthesized hardware when the same input vectors are applied to both designs. There are subtle implications of this requirement when applied to certain kinds of VHDL descriptions. We will analyze this requirement with respect to VHDL descriptions which describe clocks (sequential logic), and describe how those designs are synthesized and initialized.

A clocked process is one in which the behavior of the process is executed when a rising (or falling) transition occurs on a signal. In general, clocked processes are of the form:

```
PROCESS
BEGIN
   WAIT ON clk UNTIL clk = '0';
-- Other statements
END PROCESS;
```

In such a process, if the signal clk is a bit, whenever the value of clk changes from '1' to '0', the statements which follow the WAIT statement are executed. Note that the value '0' in the above code fragment signifies the falling edge of the signal clk. The value '1' would indicate a rising edge.

Consider the example which appears below. It contains a simple clocked process, in which the input signal (i) is inverted and assigned to the output signal (o). This assignment takes place on the falling edge of the clock (clk).

```
ENTITY exmpl IS
PORT( i, clk : IN  bit;
         reset   : IN  bit;
         o       : OUT bit );
END exmpl;
ARCHITECTURE a OF exmpl IS
BEGIN
   PROCESS
   BEGIN
      o <= NOT i;
      WAIT ON clk UNTIL clk = '0';
   END PROCESS;
END a;
```

The VHDL language has very detailed rules which specify how VHDL is simulated. The first step in VHDL simulation is elaboration. Elaboration is the process during which the values for signals at time 0.0 are calculated.

Simply stated, this can be done by setting signals to their initial values, and then executing every process until it suspends. This is how the values for all the signals in the design at time 0.0 are determined.

For the elaboration of this example, the signals i and o are set to '0' at the start of elaboration. The value is '0' because the type of i and o is bit, and the default initial value of any type is its left-most value, which is '0' for bit. When the only process in this design is executed, the assignment statement is evaluated and the value of o is set to '1'. The process then suspends when the WAIT statement is encountered because, although the value of clk is '0', the semantic of WAIT is such that the value of clk must have changed in addition to the condition being true. So the value of the signal o at time 0.0 is '1'.

The result of simulating the VHDL description above with a regular clock signal for the signal clk and an arbitrary waveform for the signal i is shown in the trace below.

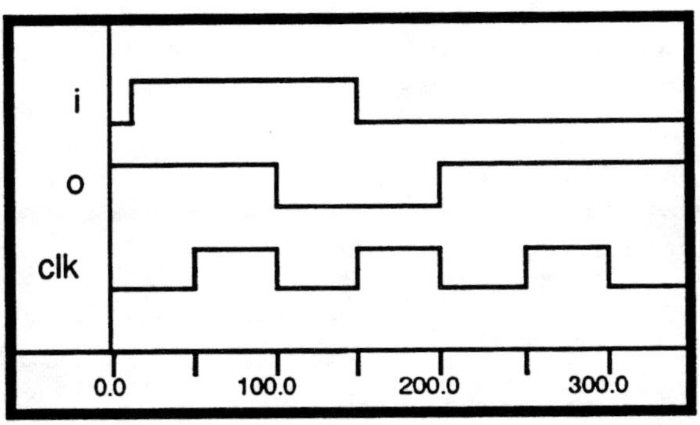

Figure 1. Simulation results of VHDL when the WAIT statement is located at the end of the process.

Now consider the example which appears below. It is very similar to the previous example, except that the WAIT statement is now the first statement in the process instead of the last.

```
ENTITY exmpl IS
   PORT( i, clk : IN  bit;
         reset  : IN  bit;
         o      : OUT bit );
END exmpl;

ARCHITECTURE a OF exmpl IS
BEGIN
   PROCESS
   BEGIN
      WAIT ON clk UNTIL clk = '0';
      o <= NOT i;
   END PROCESS;
END a;
```

When this design is elaborated, the results are different. Just as before, the signals i and o are set to '0'. When the process is executed, the process immediately suspends because the WAIT statement is the first statement encountered. So for this design, the value of the signal o at time 0.0 is '0'.

The result of simulating this example with the same input vectors as were applied to the example above is shown in the trace below.

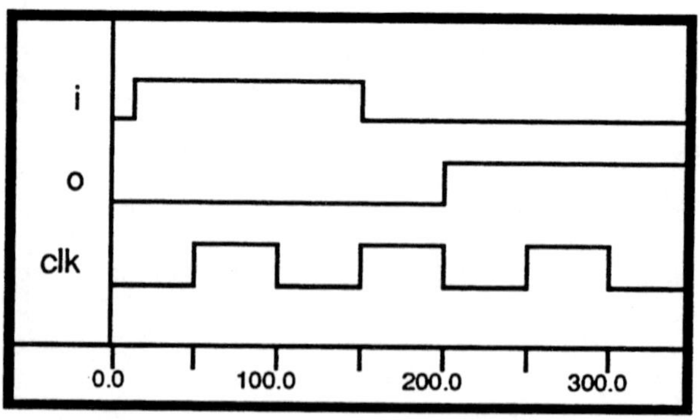

Figure 2. Simulation results of VHDL when the WAIT statement is located at the start of the process.

It is easy to see that for the first 100 ns of the simulations, the values of the signal o are not the same. This comes about because the signals were not initialized to the same value.

In order to produce correct functionality in the synthesized results, an analogy must be drawn between elaboration during VHDL simulation and the hardware implementation of the VHDL. This is done through the introduction of a global reset line.

In hardware, the initialization of circuits is usually performed using asynchronous reset lines. It is possible to use this method to mimic the elaboration which occurs at the start of VHDL simulation. A synthesis tool can insert a reset line which should be activated at the start of the hardware simulation. The process of activating this reset line will place the circuit in the same state as the result of elaboration.

A reset line is associated with a particular memory element in the hardware implementation. Thus, in order to get different implementations, when the two designs above are synthesized, the synthesis tool will associate memory elements with different signals in the design.

Memory elements are associated with signals as follows: In order for the initialization of the circuit to be correct, the "outputs" of a process must be latched when the WAIT statement is the first statement in the process and the "inputs" of the process must be latched when the WAIT statement is the last statement in the process. This can be shown with the circuits which result from synthesizing the designs.

Below is the circuit which was synthesized from the first design. In this design, the WAIT statement was the last statement in the process. As stated above, the "inputs" of the process must be latched. For this process, the only input to the process is the signal i. The implementation shows that a flip-flop was insert for the signal i. Note that the inverter follows the flip-flop, from the expression "NOT i".

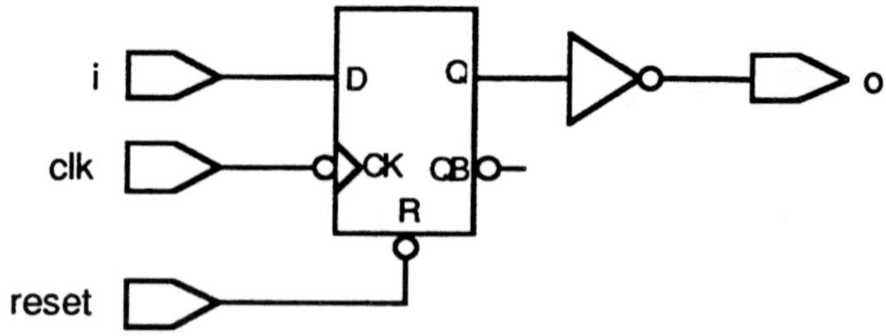

Figure 3. Synthesized result when the WAIT statement is located at the end of the process (inputs are latched).

Below is the circuit which was synthesized from the second design. In this design, the WAIT statement was the first statement in the process. Unlike the previous design, the "outputs" of the process will be latched. For this process, the only output from the process is the signal o. The implementation shows that a flip-flop was inserted for the signal o. Note that the inverter precedes the flip-flop.

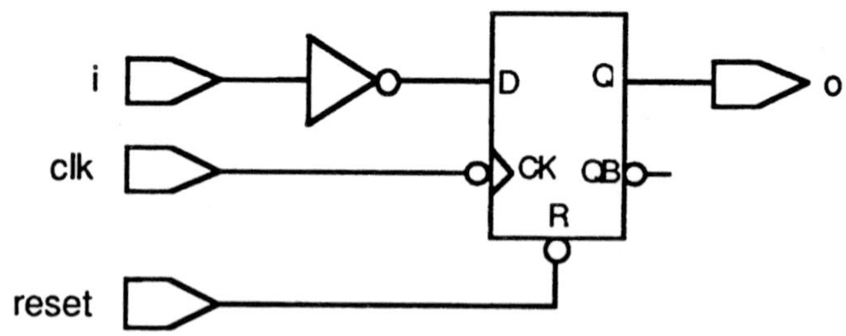

Figure 4. Synthesized result when the WAIT statement is located at the end of the process (inputs are latched).

Below is the simulation result of one of the synthesized circuits. There are two main differences to note. The first is the use of the signal reset. This is used to initialize the circuit. The value of the signal i is in an unknown state (shown by the dashed line) until this reset is complete. From this time on, the simulation is the same as that shown for the VHDL with one exception: the signal o changes state at a short time after 200.0 ns. This is caused by a technology-specific gate delay.

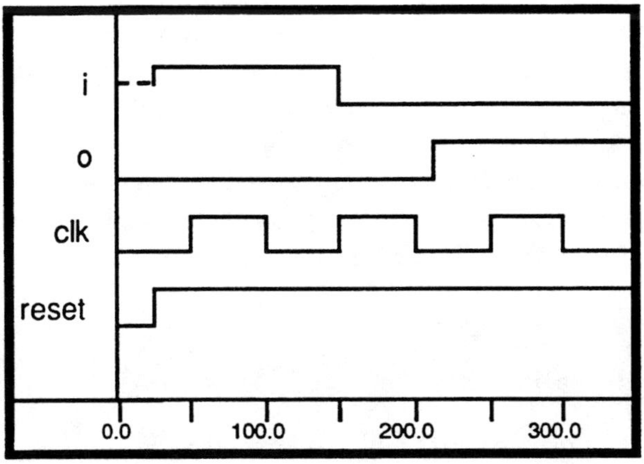

Figure 5. Simulation results of synthesized design when the WAIT statement is located at the start of the process.

We have discussed some subtleties of VHDL simulation semantics and their relationship to the synthesis process. We have shown how small changes in a VHDL description can alter the simulation semantics of a design and how our synthesis tool accurately reflects these changes in the hardware it produces.

In addition to producing designs which simulate correctly, a synthesis tool must create efficient hardware implementations for the VHDL description on which they operate. This aspect of VHDL synthesis is discussed in the following section.

Meeting Design Goals Through Optimization and Micro-Architectural Selection

One of the goals of any language based synthesis system is the production of efficient circuits. Given a description of the design written in a language, VHDL in our case, the goal is to produce a generic, technology independent, gate level implementation that is the most efficient. At the same time the synthesis system should consider the design constraints that must be met. These constraints place the target for the design somewhere along the area/speed tradeoff curve. The design may be required to meet a certain (fast) clock rate while minimizing the area cost as much as possible. Conversely, the constraint might be area while producing a clock speed that is as fast as possible.

A synthesis system can address these design goals at different points in the synthesis process. We have designed our synthesis system to apply optimizations and architectural decisions at both the language level and the gate level. The optimizations employed are geared towards the level of abstraction at which they are applied. This takes advantage of the kind of information available at the different levels. There is a set of optimizations and architectural decisions (micro-architectural selection) made at the language level which we refer to as high-level optimizations and another set of optimizations applied at the gate level. Below we discuss the language level processes.

High Level Optimizations

There are two primary goals of high level optimization. The first is to make changes to the language description that retain the behavior of the description while simplifying and reducing the size of the description (which will effect the amount of synthesized hardware). Many of these optimizations have their roots in compiler technology and so we call them compiler optimizations.

The second goal of the optimization process is to determine when operators in the language description can share a single hardware operator. This process reduces the amount of synthesized functional hardware and is called resource sharing.

Compiler optimizations.

These optimizations are geared towards simplifying the language description, and, as stated above, many of them are borrowed from compiler technology. An important aspect of these optimizations is their impact on the synthesized circuit. In some cases operators may be completely eliminated from the design. Such eliminations are based on an analysis of the language description. If these optimizations are not applied at this level of the design, it is most likely that the later gate level optimizations will not be capable of removing the unnecessary operators. This is because at the level of gates, the optimization process does not "understand" that a set of gates is performing an ADD function, nor does it understand the context of that ADD that may indicate that it can be removed.

Substantial savings in hardware can be achieved by performing these compiler optimizations.

Below we describe a number of the optimizations employed by our synthesis system.

Bit-pruning: The width of integer operators is reduced wherever possible, based on the analysis of integer range constraints. For example:

```
SIGNAL a, b : INTEGER RANGE -64 TO 63;
SIGNAL c    : INTEGER RANGE -8 TO 7;
c <= a + b;
```

Because the range of the signal c is specified as -8 to 7, only four bits are required to represent the result. The synthesized add operator is only four bits wide, rather than the 7 bits required to represent signals a and b.

Function transformation: Operators are transformed to simpler operators (whose hardware implementation is less costly) while maintaining the same function wherever possible. For example, a value that is multiplied by 2 is replaced by the value shifted left by 1.

Constant folding: Operators whose arguments are constants are evaluated and replaced by the returned value.

Constant propagation: When a variable is assigned a constant value, each subsequent use of that variable that is dependent on that assignment is replaced by the constant value.

Expression sharing: Multiple occurrences of equivalent expressions are reduced to a single expression, preventing the creation of duplicate hardware.

Expression migration: The evaluation of an expression is moved outside of a loop if the expression is not dependent on the computations of the loop. Identical expressions found in multiple clauses of an IF or a CASE statement are moved out of the IF or CASE statement.

Dead code removal: Code branches that are found to be unreachable are removed, such as when a condition in an IF statement is found to be true in all cases and only one clause is ever used.

Resource Sharing.

The goal of resource sharing is the reduction of functional hardware. Operators in the language description have the potential of being shared whenever the execution of those operators is guaranteed to never be coincident in time. For example, if an add operation occurs in each clause of a CASE statement, only one of those add operations can execute at a given time since the clauses of a CASE statement are mutually exclusive. This indicates that the add operations may all be implemented by a single adder.

Below we describe some of the resource sharing techniques employed in our synthesis system.

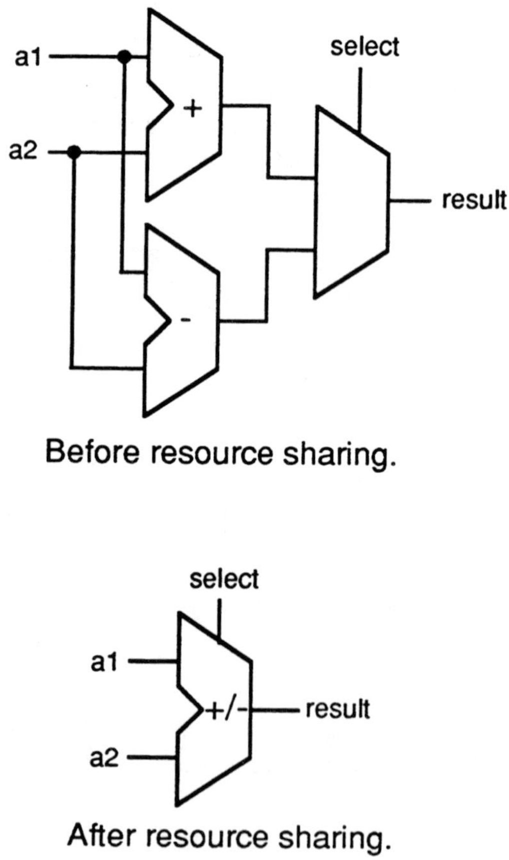

Figure 6. Separate adder and subtractor combined into an adder/subtractor.

Resource sharing: When more than one operation can be performed by a single functional unit, one functional unit is allocated instead of a unit for each operation. For example:

```
IF select = '1'
   THEN result <= a1 + a2;
   ELSE result <= a1 - a2;
END IF;
```

In this example the add and subtract operations are mutually exclusive - they will never occur simultaneously. Because of this, a single adder/subtractor can be built instead of a separate adder and subtractor. Figure 6 illustrates.

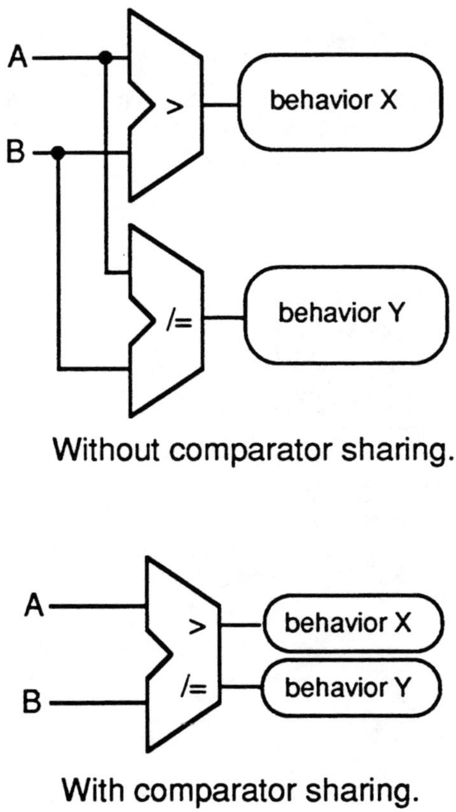

Figure 7. Comparator sharing.

Comparator sharing: Relational operations having the same operands are combined into one multi-functioned comparator. Consider the example below.

```
IF A > B THEN
      ...behavior X

IF A /= B THEN
      ...behavior Y
```

If comparator sharing was not employed, the two conditional expressions in the example would be implemented by two separator compare circuits as shown at the top of figure 7. By using comparator sharing a single comparator circuit implementing just the functions required is built as illustrated at the bottom of figure 7.

Micro-architectural Selection

The goal of our high level synthesis system is to produce a design at the technology independent gate level whose architecture targets a point along the area/speed tradeoff curve that was specified by the designer. Micro-architectural selection is the process which addresses this goal in the system. Figure 8 illustrates.

The key to micro-architectural selection is the ability to build a variety of architectures for a given operator. The choice of architecture depends on the target constraint specified by the designer.

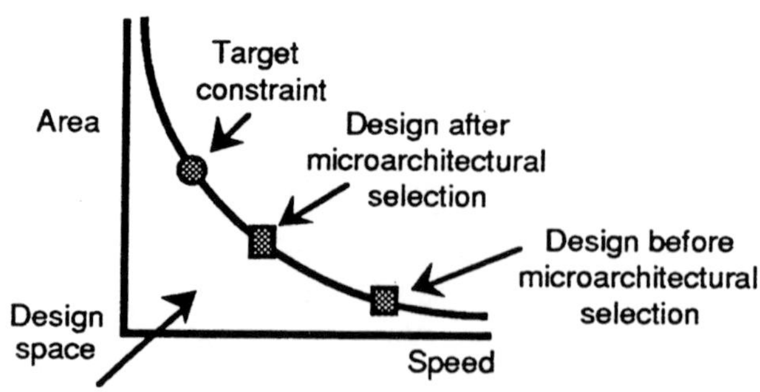

Figure 8. Micro-architectural selection moves the design implementation closer to the ultimate target, simplifying the job of logic-level optimization tools.

Consider the implementation of a 16 bit adder in the synthesized circuit. When the adder is to be built by the synthesis process, the module generator considers the target constraint. If speed is critical, the module generator will produce a highly parallel architecture for the adder. If area is the critical parameter, a full serial architecture will be produced. Depending on the constraints specified, the module generator will produce a circuit anywhere along the continuum from fully serial to highly parallel.

The result of micro-architectural selection is the creation of a technology independent circuit whose architecture targets the design goals of the designer.

By considering design constraints during high-level synthesis in the micro-architectural selection process and performing high level optimizations, our synthesis tool makes important design decisions to produce a circuit that meets design goals. Most importantly, these are decisions that can only be made at the high level, above the level of gates. By making these high level choices, the overall synthesis system does not have to rely exclusively on optimizations performed at the gate level to meet design goals. Without this ability, the low-level tools can be taxed beyond their limits, and in some cases the constraints may not be met.

Conclusion

We have discussed two important aspects of our VHDL synthesis system, AutoLogic VHDL. The first section described some of the subtleties that must be considered in the synthesis of VHDL. Small changes in the VHDL description were shown to have implications on the architecture produced by the synthesis process. These issues must be considered in order to guarantee that the functional simulation of the VHDL description is the same, on the boundary, as the simulation of the gate level design produced by the synthesis process.

In the second section we described some of the processes in place in our synthesis system to produce efficient synthesized designs. The goal of the optimizations performed is the reduction in the amount of functional hardware that must be synthesized. The process of micro-architectural selection determines the architectures of operators by considering the design goals of the user. It does this by generating implementations that occur somewhere along the serial-parallel spectrum that most closely targets the user's goals. This process moves important design decisions to the point in the synthesis system where they can be most effective, relieving the burden from lower level tools.

Part III

FORMAL VERIFICATIONS AND SEMANTICS

SYMBOLIC COMPUTATION OF HIERARCHICAL AND INTERCONNECTED FSMS

Alain DEBREIL
BULL DA, Les Clayes-sous-Bois, FRANCE
Christian BERTHET[1]
Thomson TMS, Grenoble, FRANCE
Ahmed JERRAYA
IMAG/TIM3, System Level Synthesis Group,
46, Av. Félix Viallet, 38031 Grenoble CEDEX, FRANCE

1. Introduction

In this paper we consider the problem of verifying hierarchical descriptions representing a set of interconnected synchronous Mealy finite state machines (FSM). We give a symbolic computation algorithm that builds the composite machine from the symbolic representations of the FSM components. The algorithm verifies well-formedness conditions of the resulting machine, this step detects asynchronous functional loop and bus conflicts. A tool performing this computation has been implemented and several designs has already been processed using this tool.

1.1. Motivations

To describe a complex system, hardware engineers naturally adopt a concurrent view of the hardware behavior. A complex hardware system is more easily described by a set of interconnected Finite State Machines (FSM). Traduced into an Hardware Description Language such as VHDL, a complex FSM is represented by a hierarchical structure, each leaf of the hierarchy being a component machine. This paper deals with the problem of verifying a hierarchical description representing the interconnection of several synchronous FSM.

---

[1]This research was done when the author was at BULL Research Center.

J. Mermet (ed.), VHDL for Simulation, Synthesis and Formal Proofs of Hardware, 179–193.

1.2. Objectives

The main objectives of the work described in this paper is twofold. First the verification of well-formedness conditions is performed. For example it is desirable to detect as soon as possible that the interconnection of several machines results in an asynchronous functional loop which generally means an oscillation or a hidden memory, and thus an erroneous design. Second, a symbolic computation of the composite machine; in order either to be simulated or to be formally verified by comparison with another description, which may be a functional specification or another circuit realisation.

1.3. Methodology

We start with a symbolic representation in the form of a set of Boolean functions that characterizes the behavior of each component machine. These functions represent the transition (next-state) and observable (output) functions of a synchronous Mealy FSM [Kohavi78]. We then run a symbolic computation algorithm that builds the composite machine from the symbolic representations of the FSM components. During this construction step, the algorithm verifies well-formedness conditions of the resulting machine.

Although the method presented in this paper is not directly bound to a particular representation, the algorithms have been implemented with a very efficient representation of Boolean functions, the Typed Decision Graphs (TDG) [Madre88], an improved version of the Binary Decision Diagrams (BDD) [Bryant86]. The Typed Decision Graphs already form the basis of the formal verification tools developed at BULL : PRIAM [Madre88] for the proof of combinatorial circuits and SIAM [Coudert89, Coudert90, Berthet90] for sequential circuits.

Figure 1 shows a global view of the hierarchical verification methodology.

The starting point is a hierarchical VHDL description, it will be introduced in the next section. A first step translates this format into a set of interconnected FSMs (see 2.2.2). The second step performs symbolic computation in order to check the well-formedness of the machine and to compute the composite machine that will feed the verification tools. The symbolic computation step detects Asynchronous loops and bus conflicts . The concepts and the algorithms used during this step are explained in sections 3 and 4. The two latest section provide results and conclusions.

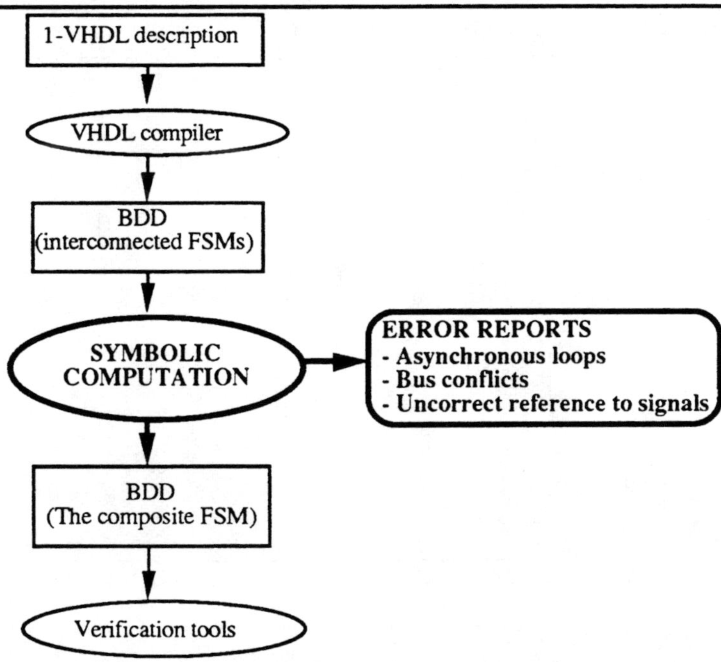

Fig1: Global view of the Hierarchical verification process

2. Specification of synchronous FSM

In the following, we use the language VHDL [Armstrong89] as a specification formalism although the method developed in this paper is not directly linked to a particular HDL. The methodology will be illustrated using the example shown in Fig 2. The device *HRR* is composed of two instances of the FSM *REGSH* (Shift Register). It may be used as a counter or as a shift register. However this machine behave correctly only under some conditions that will be explained later (section 4.4).

2.1. VHDL Hierarchical Description

A VHDL hierarchical description is a structural interconnection using component instances. In the following we restrict ourselves to a two-level hierarchy model. Besides we assume that each leaf of the hierarchy is a behavioral description corresponding to a synchronous FSM and that all the leafs of the hierarchy have the same basic clock reference (CLK in the sequel). The interconnections of the leaf entities are done either by internal signals declared in the father architecture (HRR in the example) or by the interface ports of the father entity (called interconnecting signals in the following). Figure 3 shows the VHDL description of the machine HRR. The entity HRR is composed of two instances REGSH_1 and REGSH_2 of the same component REGSH. The interconnecting signals are W1, W2, SH, DOUT0, CLK (parts of HRR interface) and BUS1 (internal signals).

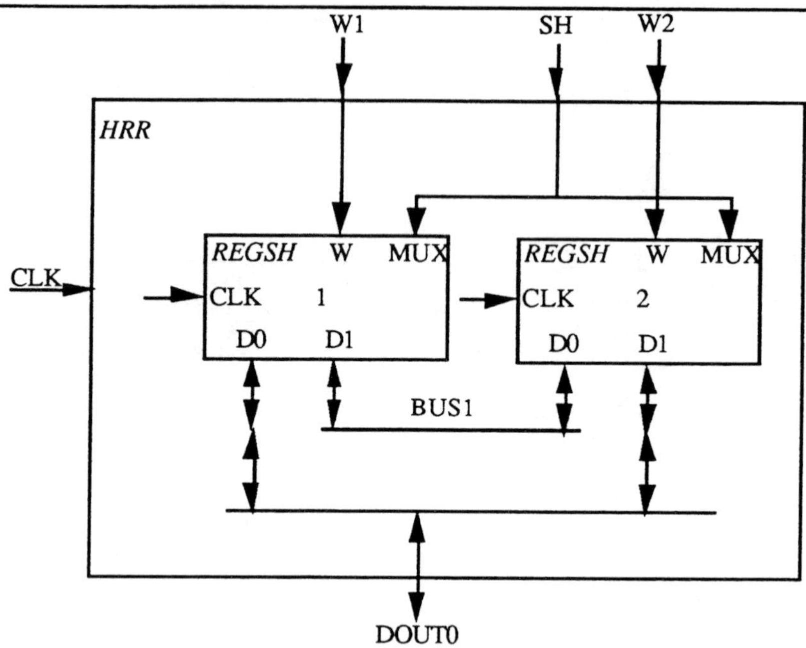

Fig. 2 Example of interconnected FSMs.

```
entity HRR is
   port(W1, W2, SH, CLK : in BIT; DOUT0 : inout busZ);
end HRR;

architecture STRUCTURE of HRR is
   component REGSH
      port(D0, D1: inout busZ,W, MUX, CLK: in bit);
   end component;
   signal BUS1 : busZ;
   begin
      REGSH_1 : REGSH port map( DOUT0, BUS1, W1, SH, CLK);
      REGSH_2 : REGSH port map( BUS1, DOUT0, W2, SH, CLK);
   end;
-- busZ is a resolved three value signal ('z' , '0' , '1')
```

Fig 3: Interconnected FSMs description example in VHDL

2.2. Behavioral VHDL Description of a Synchronous FSM

We assume the following hypotheses on the behavioral descriptions written in VHDL. The behavior of each leaf cell (i.e. a component FSM) is given using eiher the form of a VHDL process (i.e., procedural sequence of statements) at the cycle level or the form of a data-flow statement (concurrent statement). A set of VHDL processes or data-flow statements will be translated into a set of interconnected and concurrent FSMs.

Figure 4 shows the VHDL description of the component FSM REGSH. The state of the machine REGSH is encoded by a 2-bit signal (STATE) triggered by the rising edge of CLK. The output (D0) is calculated only when input W is '1' and D1 is calculated when (MUX = '0' and W = '1').

```
entity REGSH is
      port(D0, D1 : inout busZ; W, MUX, CLK : in bit);
end REGSH;

architecture BEHAVIOR of REGSH is
   signal STATE bit_vector (0 to 1);
   begin
      process (CLK, RW, MUX)
         variable STATE bit_vector (0 to 1) := "00";
         variable T : bit;
         begin
            if (MUX = '1')
                  then T := busZ_to_bit (D1);
                  else T := STATE (0 );
            end if;
            if (W = '1')
                  then D0 <= bit_to_busZ(T);
                  else D0 <= 'Z';
            end if;
            if (W = '1' ) and (MUX = '0')
                  then D1 <= bit_to_busZ(STATE(0));
                  else D1 <= 'Z';;
            end if;
            if ( CLK = '1' and not CLK'stable)
                  then STATE (0 to 1) := busZ_to_bit_vect(D0 & D1);
            end if;
         end;
   end;
-- INC is a defined function for incrementation of Bit_vectors
-- bit_to_busZ , busZ_to_bit_vect and busZ_to_bit are conversion functions.
```

Fig 4: Behavioral description example of a component FSM in VHDL

Through a symbolic execution technique [Madre88], a symbolic representation of the FSM is obtained in the form of two sets of Boolean functions : the transition functions computing the next-state (one function for each bit of the state variables) and the observable functions (one function for each bit of the out or inout ports). An output function may be uncompletely specified.

A function corresponding to an Out bus or to an Inout bus port may be uncompletely calculated. The definition context of the function represents the domain where the machine drives through the port an interconnecting signal. For example the definition context of D0 is (W = '1').

2.2.1. Representation of Uncompletely Specified Functions

An uncompletely specified Boolean function f can be represented using a pair of Boolean functions (C, V) where C is the context (or domain) of the function and V is the value on the domain [Madre88].

In the following algorithms, we use a more compact representation in the form of a single Boolean function:

$$f = (C \wedge V) \vee (\neg C \wedge s)$$

where s is a free Boolean symbol attached to the function f. For example, the function computed for the output D1 in the REGSH description is

$$\lambda_{D1} = ((W \wedge \neg\, MUX) \wedge STATE(1)) \vee ((\neg W \vee MUX) \wedge D1)$$

where D1 is a free Boolean symbol attached to the function λ_{D1}. *A function is completely calculated if and only if it does not depend on its own free symbol.*

Both representations can be derived from one another. Let f be a function and s its associated symbol. By construction of $f = (C \wedge V) \vee (\neg C \wedge s)$, the Shannon's decomposition

$$f = (f_0 \wedge \neg s) \vee (f_1 \wedge s)$$

satisfies $f_0 < f_1$. Then $V = f_0$ and $C = \neg\, \partial f / \partial s$. $\partial f / \partial s$ is the Boolean difference of f with respect to s [Thayse81]:

$$\partial f / \partial s = f_0 \oplus f_1$$

2.2.2. Representation of Component FSM

After the symbolic execution process, each component FSM is characterized by :

1- Four Bit-Vectors : In, Out, Inout, Registers. To each bit is attached a specific free *symbol*.

2- Two Vectors δ (next-state) and λ (out-and-inout) of Contexted Boolean Functions defined over the set of free symbols.

3- An initial state of the registers.

A function corresponding to an out or inout port may be uncompletely specified and then may depend on its own free symbol. The context of the function represents the domain where the machine drives through the port an interconnecting signal.

3. Symbolic computation of hierarchical descriptions

The objective is to obtain a composite (product) FSM from a set of interconnected component FSM. The composite machine has the same definition as the component machines, i.e. it is characterized first by a set a free symbols representing the global ports (in, out and inout) and the registers (Union of the component FSM Registers), second two Contexted Functional Vectors δ (next-state) and λ (Out-and-inout) defined on the set of free symbols and third an initial state of the registers.

The interconnection of several machines consists in rewriting the next state functions of all the registers and the observable functions of the global ports by eliminating the connecting signals. This is a functional substitution process which consists in replacing any occurrence of a symbol s by a function F, and is noted $(s \leftarrow F)$ in the following.

However this composition task must be performed with care. It requires the verification of well-formedness conditions on the resulting machine.

3.1. Verification of Well-Formedness Conditions

First, the bus conflicts must be detected. It must be verified that there is at most one emitter to a bus signal at any clock cycle, since a simultaneous emission generally means an erroneous situation.

Second, it must be checked that no asynchronous functional loop is created while interconnecting several Mealy - or combinatorial - machines or interconnecting an input with an output of the same machine.

Finally, since the out or inout functions of the component machines may be uncompletely specified ('Z' value), it is necessary to check that the references to an interconnecting signal is correct when it feeds another machine. In other words, any reference to a signal should be done on its definition domain ('0' , '1').

The verification of well-formedness conditions is performed when building the resulting machine The verification algorithms are explained later in the next section.

3.2. Basic Concepts

This section deals with two generic concepts used by the computation algorithm : The generic port and the generic net.

3.2.1. Generic port

Hardware Description Languages generally make use of three kinds of connection ports : In, Out and Inout. The contexted value of an In port is defined by the external world. The value of an Out or Inout port is computed by the circuit for a given context. On the other hand, it is bussed, an Inout port may define a bidirectional connection. It is used as an OUTPUT port in a given context and, otherwise, may be used as an input driven by another machine.

The three kinds of ports may be handled by the unified concept of generic port. A generic port is defined by a triple

$$(\boldsymbol{C}, \boldsymbol{V}, \boldsymbol{port\_symbol})$$

where $\boldsymbol{C}$ is a Boolean function that defines the context where the port value is internally computed. When $\boldsymbol{C}$ is 1, the port acts as an output and its value is $\boldsymbol{V}$, which is the value computed by the circuit. When $\boldsymbol{C}$ is 0, then the port acts as an input and its value is defined by ***port_symbol***, which is a symbol whose value is fixed by the external world.

The contexted value of a generic port can be expressed by the formula :

$$(\boldsymbol{C} \wedge \boldsymbol{V}) \vee (\neg \boldsymbol{C} \wedge \boldsymbol{port\_symbol}).$$

The mapping of VHDL ports to generic ports is made as follows :

1- An input is a generic port with $\boldsymbol{C} \equiv$ False. Its value is given by its symbol ***port_symbol***.

2- A completely calculated Output (Out-or-Inout) is a generic port with $\boldsymbol{C} \equiv$ True. Its value is given by the formula $\boldsymbol{V}$.

3- An uncompletely calculated Output (Out-or-Inout) is a generic port where $\boldsymbol{C}$ is the context in which the port is used as an Output. In the case of a bidirectional (Inout bus) port, $\neg\boldsymbol{C}$ is the context in which the port may be used as an Input.

In order to make the next sections easier to read, a generic port acting as an Output or Inout will be called Output. A generic port acting as an Input will be called Input.

Besides we will make a distinction between global ports (interface of the HRR entity) and the local ports (interface of the component entities).

3.2.2. Generic Net

In this section we introduce the generic net concept which is the basic interconnecting entity. A generic net is an equipotential region corresponding either to an internal signal (local net) or to a global port (global net). A generic net is represented by a specific free symbol (***net_symbol***). Each local or global port is connected to at most one net. A net may be connected to at most one global port.

The ports connected to a net act as readers and/or writers of the net. If a global or local port may affect the net value, we say that this port is a net writer. If a global or local port is affected by the net value (i.e., depends on ***net_symbol***), we say that this port is a net reader.

For example, local outputs and global inputs are net writers. Local inputs and global outputs are net readers.

4. Algorithms

This sections details the algorithms used for symbolic computation. They are given in a general format that uses only classic Boolean operations and the concepts explained above. The implementation makes use of a powerful environment for BDD computation [Madre88, Coudert89, Coudert90, Berthet90].

The computation makes use of three procedures: Composition_and_Verification, Solve-net and Verify_Correct_Reference.

The procedure *Composition_and_Verification* (Fig. 5) acts in an incremental way. It starts by building the nets corresponding to the global ports and connecting signals (from the component instances). Then the component instances are collapsed in an incremental way by merging the corresponding FSMs within the existing structure (generic nets and already collapsed FSMs). During this process, the nets are solved (see below) in order to detect bus conflicts and asynchronous functional loops. The verification of the correct references to signals is performed after the FSM collapsing is completed.

4.1. Generic net resolution algorithms

The basic operation of symbolic computation of interconnected FSM is the generic net resolution, it is performed by the procedure Solve_net (Fig 6).

During the symbolic composition of the FSMs, when all the ports connected to a given net (the set of readers and writers, defined in 3.2.2) has been processed, the Solve_net procedure is called. It performs the net resolution in three steps, more details about these steps are provided in the next two sections. After it is solved, a local net is removed in a fourth step.

```
Procedure Composition_and_Verification
; 1- Global ports connection
For each Input
   Add port_symbol to the Writers of the corresponding Net
EndFor
For each Output
   Add port_symbol to the Readers of the corresponding Net
   Add port_symbol to the Writers of the corresponding Net
EndFor

; 2- Incremental Instances connection and Net resolution
For each Instance
   Read the corresponding FSM
   Translate the FSM
   For each Input
      Add port_symbol to the Readers of the corresponding Net
      If All the Readers and the Writers of the Net are connected
                                                  Then Solve-net
   EndFor
   For each Output
      Add port_symbol to the Writers of the corresponding Net
      Add port_symbol to the Readers of the corresponding Net
      If All the Readers and the Writers of the Net are connected
                                                  Then Solve-net
   EndFor
EndFor

; 3- Correct References Verification
Verify_Correct_Reference

EndProcedure Composition_and_Verification
```

Fig 5: The Composition and Verification algorithm

4.2. Bus Conflict Detection

Bus conflict detection (Part 1 of the Procedure *Solve_net*) can be done by analyzing the set of writers of a net : A conflict may happen when more than one writer is connected to the net.

```
Procedure Solve-net   ; Nets resolution algorithm
;  The algorithm uses three global variables :
;  Bus-Conflict Initial value = False
;  AFL Initial value = False   ; Asynchronous Functional Loop
;  Incorrect_Ref Initial value = False

; (1) Computation of the resulting Net Function and Bus Conflict Detection
Context-Sum = False
λ = net_symbol ; Init resulting Net function with a free symbol
For each Writer (function λi and symbol si) of the Net
   ci = ¬(∂λi/∂si)    ; Compute the writing context
   ; Boolean Difference ∂F/∂x = F(x←0) ⊕ F(x←1)
   Local-Bus-Conflict = ci ∧ Context-Sum
   If  Local-Bus-Conflict ≠ False
      Then BUS CONFLICT DETECTED
         Bus-Conflict = Bus-Conflict ∨ Local-Bus-Conflict
   EndIf
   Context-Sum = Context-Sum ∨ ci
   λ = λi(si←λ)   ; compute resulting function of the Net
EndFor

; (2) Rename local ports
For each Reader of the Net
   Rename port_symbol by netr in all FSMs. ; netr is a free symbol
EndFor

; (3) Verify asynchronous functional loop
BD = ∂λ/∂netr
If BD ≠ False
   Then ASYNCHRONOUS FUNCTIONAL LOOP DETECTED
   AFL = AFL ∨ BD
   λ = λ(netr←0) ∧ λ(netr←1)  ; Compute new Net Function
EndIf

; (4) Functional substitution
Substitute netr by λ in All functions of FSMs, in AFL and Bus-Conflict
Remove Net (and the function λ)
EndProcedure Solve-Net
```

Fig 6: The Net Resolution algorithm

The algorithm checks that the definition domains of the writers are exclusive. If this condition is not fulfilled, then the global variable ***Bus-Conflict*** accumulates the conflict conditions. Then the net value is computed in the form

$$(\boldsymbol{C} \wedge \boldsymbol{V}) \vee (\neg \boldsymbol{C} \wedge \boldsymbol{net\_symbol}).$$

If $\boldsymbol{C}$ is not equal to True, then the net is uncompletely calculated. Note that this computation is done by iterating a functional substitution for each writer of the net, and the order of substitution does not matter insofar as the ***Bus-Conflict*** condition is False.

Extensions can be made to deal with bus resolution for multiple emissions. Besides, when an interconnecting signal is explicitly declared pulled-up (resp. pulled-down), the corresponding free symbol (***net_symbol***) is replaced by 1 (resp. 0) after bus resolution.

The two other problems are handled by analyzing the relationship between the net writers and the net readers (assuming that each reading port is rewritten by the free symbol ***netr*** as done in Part 2 of the Procedure *Solve_net*).

4.3. Asynchronous Functional Loop Resolution

An Asynchronous Functional Loop is detected by the algorithm (Part 3 of the Procedure *Solve_net*) when the writers' resulting function λ depends on the net reader symbol (***netr***), i.e. the Boolean Difference of λ with respect to ***netr*** is not equal to False.In that case we have:

$$\partial\lambda/\partial \boldsymbol{netr} = \lambda(\boldsymbol{netr} \leftarrow 0) \oplus \lambda(\boldsymbol{netr} \leftarrow 1) = \text{True}$$

When an Asynchronous Functional Loop is detected two cases should be distinguished:

1- The net acts as a hidden memorising element if:

$$\neg\lambda(\boldsymbol{netr} \leftarrow 0) \wedge \lambda(\boldsymbol{netr} \leftarrow 1) = \text{True}$$

2- The net acts as an oscillator if:

$$\lambda(\boldsymbol{netr} \leftarrow 0) \wedge \neg\lambda(\boldsymbol{netr} \leftarrow 1) = \text{True}$$

In any other case $(\lambda(\boldsymbol{netr} \leftarrow 0) \wedge \lambda(\boldsymbol{netr} \leftarrow 1))$, the net is a combinatorial function.

An oscillation or hidden memory condition must absolutely be avoided since it may result in a non-predictable situation. The global variable ***AFL*** used in the algorithm accumulates these conditions on the states and inputs. If this condition does not hold, the resulting function of the net is then:

$$\lambda = \lambda(\boldsymbol{netr} \leftarrow 0) \wedge \lambda(\boldsymbol{netr} \leftarrow 1).$$

Our model implies a zero-delay behavior in opposition to VHDL which uses δ delays (unit-delays). Thus no race (or meta-stability) condition can be taken into account. That justifies the fact that we do not handle asynchronous memorising loops.

4.4. References to Uncompletely Calculated Signals.

References to uncompletely calculated net values are detected a posteriori when the algorithm stops This operation is performed by the procedure *Verify_Correct_Reference* detailed in Fig. 7. The detection consists in verifying that any next-state or out or inout function does not depend on the symbols attached to internal signals or global out or non-bussed inout ports of the HRR interface, since in these case no information will be provided from the external world. The verification of functional dependency on the bussed out or inout ports should be carried out at the upper level.

```
Procedure Verify_Correct_Reference
For each net_symbol attached to an internal signal, global out or global
non-bussed inout
  For each function f, (λ or δ) of the resulting FSM
    BD = ∂f/∂net_symbol
    If BD ≠ False  Then USE OF AN INCOMPLETELY CALCULATED
SIGNAL
      Incorrect_Ref = Incorrect_Ref ∨ BD
    EndIf
  EndFor
EndFor
EndProcedure Verify_Correct_Reference
```

Fig 7: The Verification of Correct References algorithm

The composite machine is correct if and only if it is used in such a way that :

$$\neg AFL \wedge \neg Bus\text{-}Conflict \wedge \neg Incorrect\_Ref = \text{True}$$

The results of running the algorithms on the example machine described in section 2 detects asynchronous functional loop when SH and W1 and W2 are all high, and a bus conflict when W1 and W2 are both high and SH is low.

5. Results

A tool that performs the algorithms depicted above has been realized and tested within the chain of tools for formal verification developed at BULL, Pâris [Madre90]. Several examples have been processed. The table in Fig 8 shows the computation time for three examples.

For each example, the table provides the number of components, the number of inputs, the number of outputs, the number of registers, the number of symbols (free variables, see 2.2.2), the number of TDG nodes (optimized BDD) and the execution time obtained with a BULL workstation which performance is estimated to 50% of a SparkStation 1.

The two first rows, in the table, show the results for the two descriptions example detailed above (HRR, REGSH) in the paper. The CLM5 machine is an industrial application including a complex function using 8 communicating FSMs and 43 registers.

It expected that such a technique may be applyed successfully for the verification of complex systems. In fact the limitation of the present implementation is related to the limitation of the BDD representation we are using. For some examples, the intermediate computation induce an explosion in the size of the internal representation, the BDD. In some cases, this problem may be avoided by using a better ordering of the variables for the BDD decomposition.

Circuit	Component count	Input signal count	output signal count	registers count	Symbol count	TDG Node count	CPU Time seconds
HRR	2	4	1	4	25	64	2.4
HCR	4	4	2	4	39	40	2.7
CLM5	8	14	8	43	342	8089	1135

Fig 8: Results

6. Conclusion

We have presented a unified model for the symbolic computation of interconnected sequential machines. We detailed a set of algorithms that builds the composite machine starting from hierarchical FSM description. The algorithms verifiy well-formedness conditions of the resulting machine, they detect asynchronous functional loop and bus conflicts. A tool performing this computation has been implemented and several designs has already been processed using this tool.

This experience is used to build an industrial version of the tool that will be integrated into the chain of tools for formal verification developed at BULL [Madre90].

REFERENCES

[Armstrong89] J. R. Armstrong, *Chip-Level Modeling with VHDL*, Prentice Hall, 1989.

[Berthet90] Berthet C., Coudert O. and Madre J.C.M., "New Ideas on Symbolic Manipulations of Finite State Machines", accepted for publication ICCD, September 1990 .

[Bryant86] Bryant R.E., "Graph-Based Algorithms For Boolean Functions Manipulations", *IEEE Transactions on Computers*, Vol. C35, N.8, August 1986.

[Coudert89] O. Coudert, C. Berthet, J. C. Madre, "Verification of Sequential Machines Using Boolean Functional Vectors", *IFIP International Workshop, Applied Formal Methods for Correct VLSI Design*, Leuven, November 1989.

[Coudert90] O. Coudert, C. Berthet, J. C. Madre, "Formal Boolean Manipulations for the Verification of Sequential Machines", *EDAC Conference*, Glasgow, Mars 1990.

[Kohavi78] Z. Kovahi, *Switching and Finite Automata Theory*, McGraw- Hill Book Edition, 1978.

[Madre88] Madre J.C. and Billon J.P., "Proving Circuit Correctness using Formal Comparison between Expected and Extracted Behaviors", *Proc. of the 25th Design Automation Conference*, 1988.

[Madre90] Madre J.C., Coudert O., M. Currat, A. Debreil and C. Berthet, "The Formal Verification Chain at BULL", Proc. EUROASIC 90, Paris, France, May 1990.

[Thayse81] A. Thayse, "Boolean Calculus of Differences", LNCS 101, Spriger-Verlag, 1981.

Formal semantics of VHDL timing constructs

Ashraf Salem & Dominique Borrione
ARTEMIS-IMAG, BP-53X, 38041 Grenoble cedex, France

Abstract

The aim of the work presented here is to enlarge the subset of VHDL which can be manipulated by formal verification tools by including the timing constructs. In this paper we give formal semantics for these constructs. And, we prove, partially, the equivalence between these semantics and the informal operational semantics of the language as defined in the VHDL language reference manual. Also, we show how these semantics can establish a basis for the construction of formal timing verifiers.

1.Introduction

Most HDL-based formal verification systems are constructed around synchronous hardware description languages, where the use of the finite state machines semantics to reason about the descriptions written in these languages is justified. The same approach has been applied to VHDL by defining subsets and description styles [1] to avoid the unnecessary use of the language timing model semantics in the case of synchronous circuits. This, however, leaves out a large set of VHDL descriptions, which apply to inherently asynchronous circuits, or to the detailed levels of synchronous circuit designs.

In order to prepare for the development of methods and tools that formally verify the correct logical and timing behaviour of asynchronous VHDL descriptions, a necessary first step involves the formal definition of the appropriate VHDL primitives: signals, signal attributes, signal assignment statements with transport and inertial delays. The operational semantics of these primitives are defined in the Language Reference Manual (LRM) [2] in terms of an event driven simulation engine. This type of semantics is implementation oriented, and it is not suitable for formal reasoning.

J. Mermet (ed.), VHDL for Simulation, Synthesis and Formal Proofs of Hardware, 195–206.

We propose functional semantics for these constructs, where the semantic functions calculate the values of signals at all time points; thus, they can be manipulated formally as time functions. As a validity test of the adopted functional definitions, we used our semantic model, together with an automatic proof system, to prove as theorems a variety of properties (on signals and signal attributes) stated in the LRM.

2. The VHDL time model and our semantics

VHDL uses a two dimensional time model: the physical time and the delta delay. The physical time is the real time being described and simulated: it is discrete, with the femtosecond as the finest time division. The delta delay has been defined for the purpose of simulating a set of concurrent processes on a monoprocessor. It is used in the language to serialize the processes parallel execution. In the LRM, this serialization is called the simulation cycle. One or more simulation cycles may be necessary to obtain the correct values for the signals at a physical time instant (several cycles are necessary in the presence of zero-delay assignments to signals).

The semantics we propose here treats only the physical time. All the assignments in a source VHDL, assumed to be proved using these semantics, shall have a physical time delay, i.e an **after** construct with a non-zero time. Under this restriction, only one simulation cycle is needed at each physical time point to calculate the signal values; thus, the simulation cycle semantics can be ignored. This restriction is in accordance with the fundamental hypothesis of its application to timing verification, where all the gates used in a model have a propagation delay.

3. The Semantics of Signals

The history [3] of a VHDL signal is modelled by a sequence of values; all of these values have the same type which is the signal type. The smallest VHDL time unit, the femtosecond, represents the temporal distance between two successive elements in the sequence. The semantic domain Signal is defined as follows:

Signal = Integer* + Boolean* + ...

The signals can take values from the domain:

Value = Integer + Boolean + ...

And time is represented by the natural numbers :

Time = Natural

In our model, values in a sequence X are numbered left to right from 0 to len(X)-1, where len(X) is the length of the sequence. The current value of the signal is the head (hd) of the sequence and the signal initial value is el(len(X) - 1, X) , where function el(n,X) returns the nth element in the sequence X[4]. Thus, graphically, the current value is the leftmost element, the initial value the rigtmost element.For instance, the history of the signal X shown on Figure(1) is represented by the sequence: X = (1 1 1 0 1 0 0). The current value is hd(X) = 1 and its initial value is:

initial-value(X) = el(len(X)-1, X) = el(7-1, X) = 0.

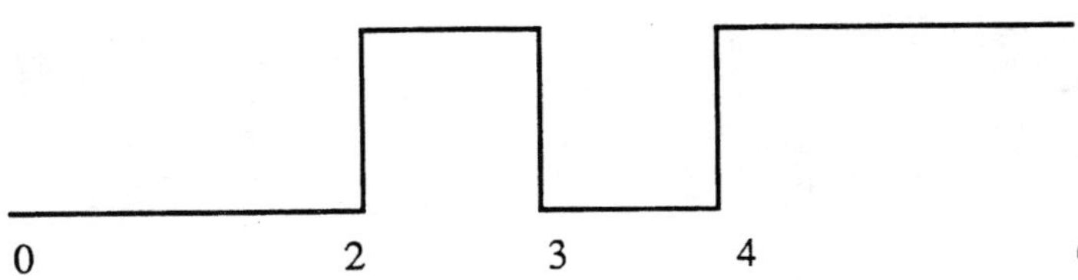

Figure(1): The waveform of the signal X

4. The Signal Attributes

Signal attributes are predefined functions taking one signal as parameter, and zero or one more parameter of a value type. They are called by prefixing the attribute name by the signal to which it applies.

4.1 The attribute delayed

S'delayed(T) is a signal S1 equivalent to S delayed T units of time. S1 is obtained from S by removing the T most recent values of S, and appending the initial value T times on the right. The following function gives the semantic of this attribute.

delayed: Signal x Time -> Signal

```
delayed = λ (s t) . if len(s) = 0 then nil
                    else  if len(s) ≤ t
                          then initial-value(s) . delayed(tl(s),t)
                          else el(t,s) . delayed(tl(s),t)
                          endif
                    endif
```

where tl(s) returns the remainder of the sequence after removing the first element and '.' adds an element before the head of the sequence.

Example:

X'delayed(3) = (0 1 0 0 0 0 0),
where X is the signal shown in Figure(1).

4.2 The attribute stable

S'stable(T) is a boolean signal which has the current value True, when S has not been changed during the previous T units of time else it has the current value False. The following function stable gives the semantics of this attribute:

Stable : Signal x Time -> Signal

```
Stable = λ (s t) .  if len(s) = 0 then nil
                    else Sta(s,t) . Stable(tl(s),t)
                    endif
```

The auxilliary function **Sta** returns the boolean value True when the signal S is stable for T units of time, and returns false otherwise.

Sta : Signal x Time -> Boolean

```
Sta = λ (s t) . if t = 0 then True
                else   if len(s) = 1 then True
                       else Sta(tl(s), t -1) and (hd(s) = hd(tl(s)))
                       endif
                endif
```

Example:

If X is the signal as of Figure (1) then
Stable(X,1) = (true true false false false true true).

4.3 The attribute last_value

The attribute last_value of S returns the value of this signal just before the last change of its value. The function last_value gives the semantic of this attribute:

Last_Value : Signal -> Value

```
Last_Value = λ s .  if len(s) = 0 then nil
                    else
                      if len(s) = 1 then hd(s)
                      else    if hd(s) = hd(tl(s)) then Last_Value(tl(s))
                              else hd(tl(s))
                              endif
                      endif
                    endif
```

Example:

X'Last_Value = 0, where X is the signal shown in Figure (1).

4.4 The attribute last_event

The attribute last_event of signal S returns the amount of time which has elapsed since the last event occured on this signal. The semantics are given by:

Last_event : Signal -> Time

```
Last_Event = λx.  if len(x) = 0 then 0
                  else
                     if len(x) = 1 then 0
                     else
                              if hd(x) ≠ hd(tl(x)) then 0
                              else Last_Event(tl(x)) + 1
                              endif
                     endif
                  endif
```

Example:

X'Last_Event = 2, where X is the signal shown in Figure(1).

5. The signal assignment statements

The presence of the after constructs in the signal assignment statements implies the use of either the inertial or the transport delay. The LRM gives preemptive semantics [5] for these two types of delay, scheduling future values which may be deleted from the projected waveform by a later execution of the assignment. Our semantics always look to the past, hence no deletion is possible.

5.1 Signal Assignment Statement with transport delay

Each signal assignment statement defines a driver for its target signal, and in the same process only one driver exists for a given signal. The semantics presented here assume a single driver for each signal, i.e the signal is affected in one process only; and in that process we assume that the signal is affected once.
Consider the following signal assignment statement:

```
P: process (X)
begin
     Y <= transport X after n fs;
end process;
```

The function Transport(x,n,y0), where y0 is the initial value of the signal y, gives the semantics of the assignment:

Transport : Signal x Time x Value -> Signal

```
Transport = λ (x n y0) .
              if len(x) = 0 then nil
              else  if len(x) ≤ n then  y0.Transport(tl(x), n, y0)
                    else  hd(delayed(x,n)).Transport(tl(x),n,y0)
                    endif
              endif
```

5.2 Signal Assignment Statement with inertial delay

The inertial delay models the transmission of the signals in the switching circuits. A signal value is transmitted from source to target only when its duration is larger than the switching time. In the LRM, the non-transmission of a signal change is expressed by a couple of actions: first the value change is scheduled, then this prevision is deleted. In our semantic model, where the full history of a signal is considered to be always available, the value of the target is computed by looking to the past events of the source.

Consider the following process statement which contains one signal assignment with inertial delay:

```
P: process(X)
begin
        Y <= X after n fs;
end  process;
```

The function Inertial(x,n,y0), where y0 is the initial value of the signal y, gives the semantic of this assignment:

Inertial : Signal x Time x Value -> Signal

```
Inertial  = λ (x n y0).
              if len (x) = 0 then nil
              else
                 if len (x) ≤ n then  y0.Inertial(tl(x),n,y0)
                 else
                    if last_event (x) ≥ n then
                     hd(delayed(x,n)).Inertial(tl(x), n ,y0)
                    else
                       if last_event(delayed(x,last_event (x))) ≥ n
                       then hd(delayed(x,n)).Inertial(tl(x),n,y0)
                       else  hd(Inertial(tl(x),n,y0)).Inertial(tl(x),n,y0)
                       endif
                    endif
                 endif
              endif
```

The timing diagram of Figure(2) shows the various cases of this function using the following example: Y <= X **after** 2 **fs**;

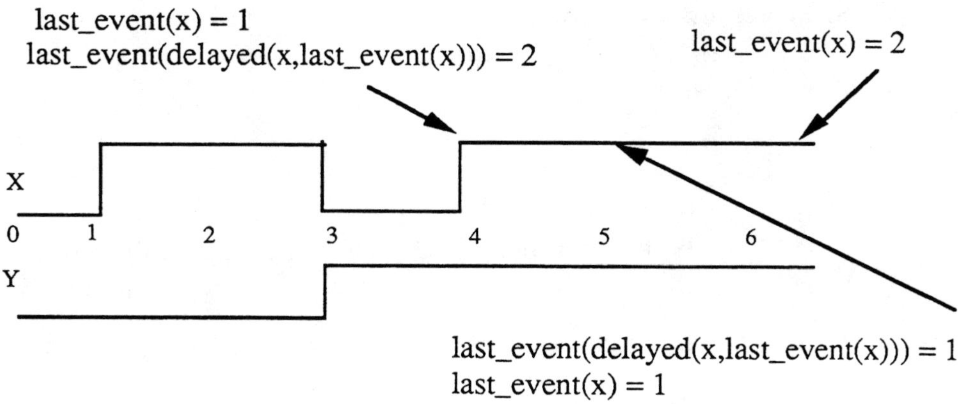

Figure(2): The inertial delay characteristics

6.Validation of the VHDL functional semantics against its operational one

The LRM states, in plain English, a number of relations among the different signal attributes, as a means for the reader to understand better the attributes definition. We consider that proving these relations as lemmas within our semantics constitutes a partial validation of our model. To this aim, we wrote our semantic definition for the VHDL signals, signal attributes, and delayed assignment statements in the Boyer-Moore's logic, and submitted these relations as lemmas which were shown correct use the Boyer-Moore theorem prover [6].

Lemma 1:

"If Ts is the smallest value such that S'STABLE(Ts) is FALSE, then for all t where 0 ns ≤ t <Ts, S'DELAYED(t)=S " (LRM, page 14-8).

We express this lemma by the following first order predicate:

∀ t, t > 0 & t ≤ ts & signal(s) & hd(stable(s,ts) & not(hd(stable(s,ts+1)))
=> hd(delayed(s,t) = hd(s)

Lemma 2:

"If S'STABLE(Ts) is FALSE, then by definition for some t where 0ns ≤ t ≤ Ts , S'DELAYED(t) ≠ S" (LRM, page 14-8).

This lemma can be transformed to the following one: If, for all t such that 0ns ≤ t ≤ Ts, S'DELAYED(t) = S then S'STABLE(T) is TRUE.

Which we express as follows:

∀ t, t > 0 & t ≤ ts & signal(s) & hd(delayed(s,t) = hd(s)
=> hd(stable(s,ts))

Lemma 3:

"For a scalar signal S, S'LAST_VALUE = S'delayed(T) where T ≥ 0 ns is the smallest value such that S'Stable(T) is False" (LRM, page 14-8).

This lemma can be expressed formaly using the following first order expression:

∀ t, t>0 & signal(s) & hd(stable(s,t)) & not hd(stable(s,t+1))
=> last_value(s) = hd (delayed(s,t+1))

Lemma 4:

"For a scalar signal S, S'LAST_VALUE = S'delayed(T) where T ≥ 0 ns is the smallest value such that S'Stable(T) is False,If no such T exists, then S'LAST_VALUE is equal to S" (LRM, page 14-8).

This lemma can be modelled in our semantics as follows:

∀ t, t > 0 & signal (s) & true(stable(s,t)) => hd(s) = last_value(s)

Lemma 5:

"For a scalar signal S, S'LAST_EVENT returns the largest value T of type time for which S'DELAYED(T)'STABLE would return TRUE "(LRM, page 14-7).

The lemma can be expressed by the following first order perdicate:

```
signal (s) & last_event (s) = t
        =>      hd (delayed ( stable (s, 1),t)) &
                not(hd(delayed(stable(s,1),t+1)))
```

Lemma 6:

"Let R be of the same subtype as S, let T ≥ 0ns, and let P be a process statement of the form:

```
P: process(S)
   begin
      R<= transport S after T;
   end process;
```

Assuming that the initial value of R is the same as the initial value of S, then the attribute 'delayed is defined such that S'delayed(T) = R for any T." (LRM, page 14-5).

The lemma can be expressed as follows:

∀ t, t>0 & signal(s) & r0 = initial-value(s)
=> delayed (s, t) = Transport(s, t, r0)

7. The semantics and the timing verifiers

Timing verification can be done using a term-rewriting based tool [7], performing timing-property preserving transformations on the HDL constructs. A number of rewriting rules were proposed in [7, 8] to do such transformations. To justify the use of these rules in building a VHDL formal timing verifier, their validity should be proven in the language semantics model. In this paper, we consider some of these rules, which have been proven correct with the Boyer-Moore theorem prover, and we show how they can be formulated in our semantics.

Lemma 7:

Let S be a scalar signal and let m ≥ 0 and n ≥ 0, then S'delayed(n)'delayed(m) = S'delayed(m+n) [7].

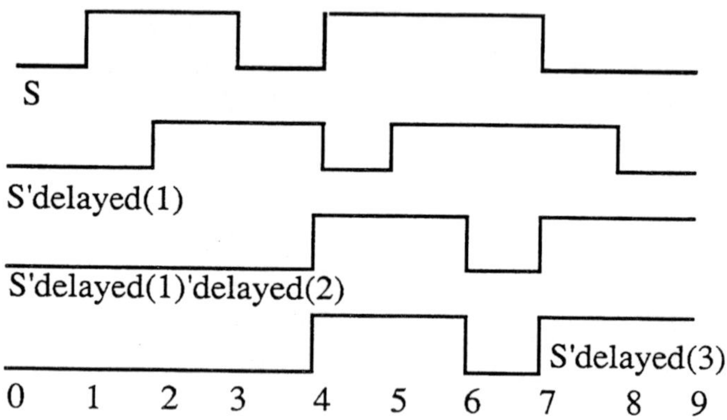

This lemma is written as:

$$m \geq 0 \ \& \ n \geq 0 \ \& \ \text{signal}(s) \Rightarrow \text{delayed}(\text{delayed}(s,m),n) = \text{delayed}(s,n+m)$$

Lemma 8:

Let S be a scalar signal and let 0≤m and 0≤n then S'delayed(n)'stable(m) = S'stable(m)'delayed(n) [7].

This lemma can be expressed as follows:

$$n \geq 0 \ \& \ m \geq 0 \ \& \ \text{signal}(s) \Rightarrow \text{hd}(\text{delayed}(\text{stable}(s, m), n)) = \text{hd}(\text{stable}(\text{delayed}(s, n), m))$$

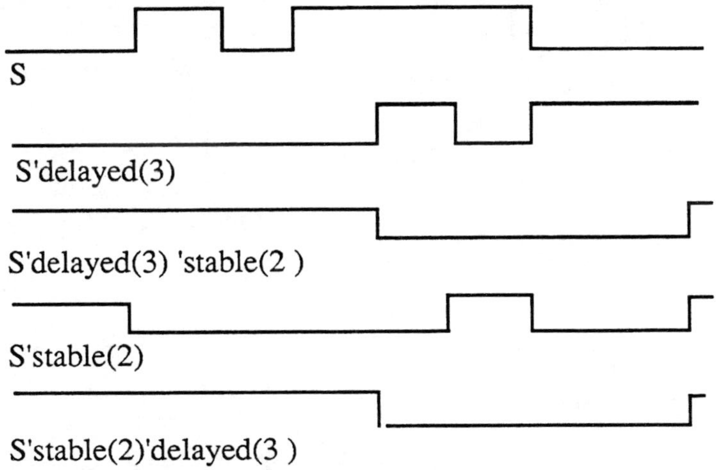

Figure(3): S'stable(n)'delayed(m) = s'stable(m)'delayed(n)

Lemma 9:

Let R be of the same subtype as S, let $n \geq 0$ ns, and let P be a process statement of the form:

```
P: process(S)
begin
    R<= S after n;
end process;
```

Assuming that S'stable(m) is true where $n \leq m$, then S'delayed(n) = R.

This lemma can be expressed as follows:

signal(s) & n≥0 & m≥0 & hd(stable(s,m)) & n ≤ m
=> hd(delayed(s,n)) = hd(inertial (s, n, r0))

8. Conclusions

We defined functional semantics for the VHDL timing constructs. These semantics are not meant to be translation semantics, i.e not to be used as a target logic by a proof-oriented compiler to generate provable code from a source VHDL description. Instead, they establish a formal basis to demonstrate the temporal properties usually taken as axioms, and directly implemented in a formal timing verifier.

The validity of these semantics with respect to the language operational semantics, as given in the LRM, is partially established by the formal proof, within the semantic model, of the signal attributes properties stated in chapter 14 of the LRM.

Further work aims at removing some simplifying hypothesis, and defining the semantics of the VHDL timing constructs in the general case, including: sequences of assignments to the same signal in a process, multiple drivers to a signal, resolution functions, conditional waveforms, etc... We shall also enlarge the set of rewriting rules which may be proved using these semantics.

Acknowledgements

This work was partially supported by the EEC under contract ESPRIT BRA 3216 "CHARME", and by CNET-CNS under contract N°903B081007909245CNS

References

[1] Dominique Borrione, Laurence Pierre, Ashraf Salem: "PREVAIL: A Proof Environment for VHDL descriptions", Advanced Research Workshop on Correct Hardware Design Methodologies, Turin, Italy, June 1991.

[2] IEEE : "IEEE Standard VHDL Language Reference Manual", IEEE press, 1988.

[3] G.Kahn: "The semantics of a simple language for parallel programming", Information processing 74, North-Holland publishing Company.

[4] M.C.Gordon, "The denotational description of programming languages", Springer-Verlag, 1979.

[5] Lary M. Augustin: "Timing Models in VAL/VHDL", Proc. ICCAD'89, Santa Clara, Nov 1989, pp. 122-125.

[6] R.S Boyer and J.S. Moore: "A computational logic handbook", Academic Press, 1988.

[7] Hans Eveking and Christoph Mai : "Formal Verfication of Timing Conditions". Proc. EDAC'90, Glasgow, 12-15 March 1990, pp 512-517.

[8] J.Barros and B. Johnson: "Equivalence of the Arbiter, the Synchronizer, the Latch, and the Inertial Delay", IEEE-TC, July 1983.

A Structural Information Model of VHDL

R. A. J. MARSHALL and H. J. KAHN
University of Manchester, Oxford Road, Manchester, M13 9PL, U.K.
Phone : +44 61-275-6289, Fax: +44 61-275-6280
Email: rmarshall,hkahn@cs.man.ac.uk

Abstract. This paper describes the production of a preliminary version of an information model of VHDL. Such a model may be used to clarify the relationship between VHDL constructs and those in other CAD systems and intermediate formats. It may be used to supplement the Language Reference Manual to assist understanding of VHDL capabilities and semantics. This understanding will also aid the design of interfaces between VHDL and other formats for CAD data.

The modelling is being carried out using the information modelling language EXPRESS. This language enables a domain of interest to be defined in terms of a set of objects, relationships and constraints. The IEEE standard specification of VHDL 1076, is used as the source of the information model.

The purpose of producing an information model of VHDL is that such a model will enable the investigation of the internal consistency of VHDL and of its consistency with other CAD formats. The information model being developed for VHDL holds information not only on the use of VHDL to describe the structure and behaviour of the object being modelled, but also on the semantic restrictions contained in the reference manual. For example, a structural VHDL description allows architectures to contain internal signals and component instances, whereas, the semantics require that the mode of an actual port should be consistent with the mode of its corresponding formal port.

This paper presents a preliminary model of the connectivity subset of VHDL, comments upon its interesting aspects and highlights the means by which a complete VHDL model may be realised.

Key words: VHDL - Information Modelling - EXPRESS

1. Introduction

This paper describes a preliminary version of an information model of VHDL. An information model of VHDL should specify the semantics of objects described by the language with respect to the particular domain of interest. Therefore an information model of the use of VHDL in synthesis would be a different model from that specifying the simulation domain. The particular domain and background of the current model is that of structural design.

The production of such a model is desirable in order that the relationship between VHDL objects and those in other CAD systems and design representations may be understood. This will facilitate a common understanding of CAD data and should aid the design of interfaces between VHDL and other CAD formats. For example, the existence of information models for EDIF[1; 4] and for VHDL will assist the comparison of the two representations and aid the production of software which need to use both EDIF and VHDL.

2. Modelling Methodology

The modelling is being carried out using the information modelling language EXPRESS[3]. This language enables a domain to be modelled in terms of a set of

J. Mermet (ed.), VHDL for Simulation, Synthesis and Formal Proofs of Hardware, 207–225.

entities[1], relationships and constraints on these entities and relationships.

Entities are collected into schemas and possess attributes which are typed. Attributes may be simple or aggregates of other entities either from the same schema or from other schemas, provided that they are visible to the entity. Attributes can also be derived. In other words, they are computed using the values of other attributes belonging to that entity. Constraints can be imposed either on an instance of an entity, or globally for all instances of that entity.

The IEEE standard specification of VHDL 1076, the Language Reference Manual (LRM)[2], is used as the source of the information model.The full information model should use not only the implied structure of the CAD objects defined by the language, but also enforces the semantic restrictions specified in the LRM relating to both the static design unit and the simulation semantics.

3. Modelling Objectives

The intention of the modelling process is to define the area of the CAD domain encapsulated by the facilities of the VHDL language. Such a model will enable the investigation of the internal consistency of VHDL and its compatibility with other CAD formats. The model should hold information not only on the structure and behaviour which can be described in VHDL, but also on the semantic restrictions contained in the LRM. To illustrate the differences, consider the following case: a structural VHDL description allows architectures to contain internal signals and component instances; in addition the semantics also require that the mode of an actual port should be consistent with the mode of its corresponding formal port. Both classes of information must be modelled. If this information model definition is correct it should be possible then to derive (and justify) the VHDL syntax from the model. It would of course also be possible to derive alternative syntaxes for VHDL from the same information model.

The preliminary version of the model, which is an extension of the work described in [5], has restricted itself to the VHDL connectivity model. This is a static view of VHDL related most closely to the information contained in a design database and therefore the actual elaboration and simulation processes have been excluded. This has been done both as a delimiter for an initial subset and because the information modelling of behaviour requires further research. In addition, this subset was chosen because it is the area which has the largest intersection with other standards. Much work has been carried out on a connectivity information model for EDIF and a direct comparison between the two formats would greatly assist inter-format communication.

The modelling of VHDL has enabled analysis of the exact relationship between the *explicit* constructs and the *implicit* semantics in the manual. These often relate

[1] The fact that VHDL uses the keyword entity to describe a unit of design and EXPRESS uses the term entity as a fundamental unit of the information model may cause some confusion. Where such confusion is possible this paper uses an emboldened **entity** for the VHDL concept.

to CAD entities which are only partly stated in the syntax of the language. The information model also makes explicit the requirements of the CAD model of VHDL which are presently not directly apparent from the syntax of the language.

Care has been taken in producing the model so that it is the *meaning* of the VHDL that is described rather than its pure syntax. This has meant that various VHDL constructs and restrictions, which are related to language consistency issues but which have no semantic CAD content, are not included.

4. Model Highlights

An abbreviated version of the model is contained in Appendix B. The following areas of the model are described here:

- the design unit hierarchy - those design units described in the LRM;
- component instantiation - the modelling of instantiating a VHDL design entity;
- the configuration process - the binding between an abstract view of a component and its implementation.

These areas were chosen because they illustrate the differences and tensions between the structure derived from the modelling process and the implied constraints imposed by the syntax and semantics of VHDL.

4.1. Design entity hierarchy

EXPRESS allows the definition of SUB and SUPERTYPEs which permit the definition of inheritance used in many object-orientated languages[2]. Figure 1 shows, in EXPRESS-G notation, the common hierarchy of the VHDL **entity**, **component** and **block** entities. The corresponding EXPRESS language fragments are also included. Attributes are depicted by single thickness lines, sub and supertypes ('is-a' relationships) by double-thickness lines. The figure excludes various parts of the model which are not directly related to **entity** declarations. This section of the model hierarchy contains the entities `library_unit`, `unit_definition`, `entity_definition`, `component_template`, `block` and `template`.

A `library_unit` is the equivalent of the VHDL LRM `library_unit`. The `library_unit` has two attributes:

- `context` - this will describe those external libraries to be made visible to the design unit;
- `source_file` - this is the name of the file containing the design unit and may be used by a debugger.

The `unit_definition` is a SUBTYPE of `library_unit` and is the equivalent of a `package_declaration` or `entity_declaration` in the LRM. This entity

[2] A brief explanation of some of the EXPRESS syntax contained in these examples is presented in Appendix A.

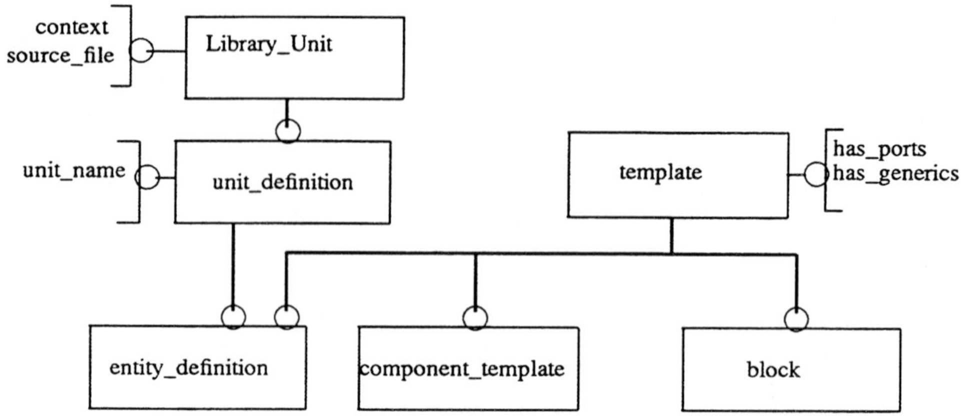

Fig. 1. The design unit hierarchy

contains the additional attribute of a name as the `unit_definition` may be accessed externally. The `unit_name` must be unique within a VHDL library. This is achieved in the information model by specifying that in all `unit_definitions` the combination of name and containing library is unique. This implies that there can only exist one unit in the given library with that name.

The other part of the hierarchy of which `entity_definition` is a SUBTYPE contains a single element - `template`. This encapsulates the interface between a VHDL **entity**, **component** or **block** and its environment. These interfaces will have common attributes which are modelled in the EXPRESS of Figure 3.

The principal differences between this hierarchy and that of the VHDL syntax are the addition of a source file and the factoring out of the `template` entity. The source file attribute is inherent in a VHDL design library. The `template` entity is a natural development of the elaboration process when an instantiated component is deemed to be equivalent to an embedded block. This equivalence will be made explicit when the configuration process is modelled; in the VHDL LRM these mappings are not apparent from the syntax.

4.2. The Relationship between Specification and Implementation

According to the VHDL reference manual, a VHDL **entity** is independent of its architectures and should have no knowledge of them. However the LRM also states that:

> 'A secondary unit is potentially affected by a change in its corresponding library unit. If a library unit is changed (e.g. by reanalysis of the corresponding design unit), then all library units that are potentially affected by such a change become obsolete and must be reanalyzed before they can be used again.'[3]

[3] VHDL LRM 11-4

```
ENTITY library;
  library_name : Name;
  has_unit_definitions: SET [0 : ?] OF unit_definition;
  has_entity_implementations:
   SET [0 : ?] OF entity_implementation;
  has_package_implementations:
   SET [0 : ?] OF package_implementation;
  has_designs :  SET [0 : ?] OF configuration;
END_ENTITY;
ENTITY Library_Unit
ABSTRACT SUPERTYPE OF (
ONEOF(unit_definition, entity_implementation,
      package_implementation, configuration));

-- packages whose resources it uses
    context : SET [1 : ?] OF unit_definition;
    source_file : OPTIONAL STRING;
END_ENTITY;
ENTITY unit_definition
    SUPERTYPE OF (ONEOF(entity_definition,package_definition))
    SUBTYPE OF ( Library_Unit);
    unit_name : Name;
DERIVE
    owning_library : library :=
USEDIN(Library_Unit,'library.has_unit_definitions')[1];
-- there can only be one member of the USEDIN aggregate
UNIQUE
unit_name,owning_library;
END_ENTITY;
```

Fig. 2. Library Structure

Thus, even though there is no explicit ownership, one cannot access the architectures after a change to the **entity**, as they are no longer implementations for the new **entity**. In EXPRESS, this functionality is modelled by the DERIVEd attribute and UNIQUEness clauses of the `unit_definition` entity. The `unit_definition` has an attribute specifying the library to which it belongs. In addition, the condition that there can only be one `unit_definition` with the given name in the library is imposed. This specification has the effect of making the implementations unavailable when the `unit_definition` is re-inserted in the library.

```
ENTITY template
ABSTRACT SUPERTYPE OF (
ONEOF(entity_definition,component_template,block));
  has_ports :   OPTIONAL SET [1 : ?] OF port;
  -- interface of template to the world
  has_generics : OPTIONAL SET [1 : ?] OF generic_parameter;
  -- parameters to be passed to the template
END_ENTITY;
```

Fig. 3. The common interface

The model of an **entity** is shown in Figure 4. This definition is an extension of the design unit hierarchy of section 4.1 and shows some of the attributes of an EXPRESS definition of a VHDL **entity**. The interface is (as previously mentioned) modelled in the `template` entity.

The `entity_definition` contains five additional attributes:

- `has_implementations` - this makes explicit the ownership by an **entity** of its implementations. This is hidden by the VHDL syntax. This is defined in EXPRESS as a list because the order of creation of the implementations is important. If a default binding indication is used, the default entity aspect is of the form

  entity *entity*_name (*architecture*_identifier)
  'where the entity name is the simple name of the instantiated component, and the architecture identifier is the same as the simple name of the *most recently analyzed* architecture body associated with the entity declaration.'[4];

- `has_types` - this is an aggregate of `VHDL_TYPE`, which is an entity defined in another schema.
- `has_signals`- this attribute represents the signals defined in the `entity_definition`.
- `local_values` - this has not been further modelled, but will represent VHDL constants used by the **entity**;
- `has_checks` - this aggregate models the *entity_statement_part* specifying the required design constraints upon the object.For example, a certain port may not be allowed to change within a given time after the device has been clocked.

The `entity_definition` has a further subtype, `simulatable_entity`. The justification for this further hierarchy is that an **entity** with no architectures is a different entity from one which possesses one or more architectures. Either entity

[4] VHDL LRM 5-6, authors' emphasis

```
ENTITY entity_definition
SUPERTYPE OF (simulatable_entity)
SUBTYPE OF ( unit_definition,template, Instantiable_Unit);
    has_implementations : LIST [0 : ?] OF entity_implementation;

-- Definitions used by the design unit to define types,
-- variables, signals used by the statements
-- private definitions  of (for example) the contents of a ROM
  has_types       :    OPTIONAL SET [1 : ?] OF VHDL_TYPE;
  has_signals     :    OPTIONAL SET [1 : ?] OF signal;
  local_values    :    OPTIONAL SET [1 : ?] OF undefined;

-- Statements used by the design unit to check it is
-- being used within its design constraints. They will involve
-- timing and modelling changes in signal values.
  has_checks : OPTIONAL SET [1:?] OF passive_process;

END_ENTITY;
```

Fig. 4. EXPRESS definition of a VHDL **entity**

may be referenced by a component configuration, but when the actual configuration takes place the reference must be to a `simulatable_entity`.

5. Comments on the Information Modelling Process

The production of the VHDL information model has enabled analysis to be performed on the relationships between VHDL semantic constructs which may have very different syntactic realisations. The close relationship between component instantiations and blocks is explicit in the EXPRESS description and is necessary in order enable the modelling of the elaboration process. The modelling has forced the authors to decide which constraints imposed by the LRM are 'real constraints' and which are merely syntactic requirements.

The major shortcoming in the present version of EXPRESS, which has affected the modelling of VHDL, has been the lack of any means of describing the concept of consequence. A signal assignment can be modelled as a 'data structure' but there is no means of describing what it *does*. Figure 5 is a data structure definition of a VHDL signal assignment. Two local rules are shown for the signal assignment. The first rule specifies that the signal type of the waveform and the target should correspond. The second rule, which describes the 'effect' of the entity, cannot currently be modelled.

```
ENTITY signal_assignment ABSTRACT SUPERTYPE OF
 (ONEOF(transport_assignment, inertial_assignment));
  signal_target : target;
  waveforms : LIST [1:?] OF waveform_element;
  prediction_time : time; -- absolute time prediction was made
WHERE
type_compatibility : not_yet_defined;
-- the signal type of a waveform must correspond to that of
-- the target (the function body has not yet been defined)

-- ******************************
-- waveforms are applied to target
-- ******************************
END_ENTITY;

ENTITY waveform_element;
  has_time_expression : OPTIONAL name_exp;
  -- if not present then a delta delay
  has_value_exp : name_exp;
  -- if 'null' then driver no longer contributes to target
END_ENTITY;
```

Fig. 5. VHDL Signal Assignment Model

A possible solution to the problem of modelling the consequence could be as in Figure 6. In this model the target is made a subtype of signal and is explicitly associated with the values that are to drive it. However, this model would have does not yet solve the handling of the driving of a signal from multiple processes. The solution here may require an approach similar to the VHDL resolution function.

A second example of the need for consequence is the configuration process. Before elaboration, an `entity_implementation` (a VHDL architecture) may (potentially) instantiate any **entity**. If the elaboration is performed successfully then the component instance is replaced in the hierarchy by the actual entity and an `entity_implementation`. However if, when elaboration takes place, the instantiated **entity** proves to have no implementations (possibly through recompilation of the **entity**), then the elaboration, and thus simulation, cannot be performed.

Research is presently continuing on the extensions required to EXPRESS in order to model such consequences.

```
ENTITY signal_assignment ABSTRACT SUPERTYPE OF
 (ONEOF(transport_assignment, inertial_assignment));
  signal_target : target;
  prediction_time : time; -- absolute time prediction was made
WHERE
type_compatibility : not_yet_defined;
END_ENTITY;
ENTITY target SUBTYPE OF (signal);
  is_affected_by : LIST [1:?] OF waveform_element;
END_ENTITY;
ENTITY waveform_element;
  has_time_expression : OPTIONAL name_exp;
  has_value_exp : name_exp;
END_ENTITY;
```

Fig. 6. Alternative VHDL Signal Assignment Model

6. Future Work

The model described in this paper is only a preliminary version. It would be desirable to use the model both as a means of understanding the concepts implied by the VHDL LRM and as a starting point for a full version of the information model.

6.1. Preparation of the Full Model

It is intended to produce a full model of the information contained in the VHDL simulation domain. This model will have to deal with the following additional areas:

- Behavioural constructs - the present model only describes the structure of design units. This will need to be enhanced to cover both signal assignment and the modelling of timing constraints;
- Elaboration - the present model is of a static VHDL database. The full model will describe the process of elaboration produced by selecting a particular implementation of an **entity** for simulation;
- Simulation - finally the actual simulation process needs to be modelled. This will involve modelling the VHDL understanding of the simulation process as described in the LRM and the possible creation and deletion of entities.

6.2. Generation of Interfaces

It is intended that the information model be used to support the generation of programming interfaces to the design schema. This will enable the storage and recovery of instances of the schema as well as interrogation and manipulation of the entities. Constructors (entity initialisation functions) could also be written to load the schema instance from a third party VHDL intermediate form.

7. Conclusions

The initial model has highlighted some of the real objects described by VHDL which are hidden by the syntax of the language. Further work still needs to be done on the extent to which EXPRESS can support the VHDL model of behaviour or whether enhancements need to be made to the language so that the full range of VHDL objects can be modelled.

Acknowledgements

This work is being supported by the CEC ESPRIT programme as part of ECIP Project 2072 (ECIP 2) and constitutes part of Work Package 1. The support of the Commission of the European Communities is greatly appreciated. We would also like to thank other members of the CAD Group in the Department of Computer Science, particularly Alan Williams and Rachel Lau, for their support and advice.

References

1. Electronic Industries Association. *Electronic Design Interchange Format Version 2 0 0 ANSI/EIA-548-1988, ISBN 0-7908-0000-4*, March 1988.
2. Institute of Electrical and Electronics Engineers, 345 East 47th Street, New York, USA. *IEEE Standard VHDL Language Reference Manual*, 1988. IEEE Std 1076-1987.
3. ISO Technical Committee 184 (ISO TC 184/SC4/WG5/P3). *EXPRESS Language Reference Manual N14*, 1991.
4. Rachel Lau. Proposal for an information model for EDIF. Technical Report Series UMCS-91-6-2, Department of Computer Science, University of Manchester, Manchester M13 9PL, England, June 1991.
5. R. A. J. Marshall and H. J. Kahn. An Information Modelling Approach to VHDL. In *Proceedings of the First European Conference on VHDL Methods*, IMT, Marseille, September 1990.

Appendix

A. EXPRESS - a language summary

The following is a brief description of the features of EXPRESS used in the paper. For further and fuller details consult the EXPRESS reference manual[3].

EXPRESS constructs which are 'near English' (e.g. LIST [1:5] OF) are regarded as self-explanatory.

- EXPRESS remarks are initiated by two consecutive dashes.
- an ENTITY declaration may specify (apart from the entity name) :
    1. Those entities from which it inherits attributes or which inherit attributes from it (SUPERTYPE/SUBTYPE)
    2. The attributes of the current entity

3. Derived attributes (DERIVE clause)
4. Any attributes which are unique for all instances of the entity (e.g. names within a library)
   (UNIQUE clause)
5. local rules (WHERE clause)
6. Whether an entity may only be instantiated via its subtypes (ABSTRACT SUPERTYPE)

All of the above sections are optional.

- Possible EXPRESS aggregates are ARRAY, BAG, LIST and SET as well as the generalised AGGREGATE type.
- An attribute can be retyped to allow a more specific use of an attribute in an inheritance relationship. For example all `port_instantiations` have an `this_port` attribute. When `port_instantiation` is inherited by `driven_port_instantiation`, it is retyped as an `input_port`.

The EXPRESS reference manual and other STEP documentation may be obtained from NIST, Gaithersburg, USA.

B. Abbreviated EXPRESS Model

The model in this paper is an outline model for the purposes of ease of understanding. Many of the rules are only sketched with a descriptive label and a constant boolean value. The function bodies have also been omitted.

```
SCHEMA vhdl;
— This is the main schema for the VHDL information model.
— It contains the major elements of the VHDL library system, it will also
— deal with (in the future) details of VHDL elaboration and the VHDL simulator
REFERENCE FROM structural_vhdl (component_template, block,port,
  generic_parameter, signal, for_generate, if_generate,
  instance_configuration, component_instantiation);
REFERENCE FROM behavioural_vhdl(behavioural_implementation, action,
  passive_process);

TYPE
   Meaningful_Name = STRING;
END_TYPE;
TYPE
   Undefined = INTEGER;
END_TYPE;

ENTITY library;
   library_name : Meaningful_Name;
   has_unit_definitions: SET [0 : ?] OF unit_definition;
   has_entity_implementations: SET [0 : ?] OF entity_implementation;
```

```
    has_package_implementations: SET [0 : ?] OF package_implementation;
    has_designs : SET [0 : ?] OF configuration;
END_ENTITY;

ENTITY Library_Unit ABSTRACT SUPERTYPE OF
 (ONEOF(unit_definition, entity_implementation,
   package_implementation, configuration));
— packages whose resources are visible (STANDARD is always present)
   context :  SET [1 : ?] OF package_definition;
   originator : OPTIONAL STRING;
END_ENTITY;

ENTITY Instantiable_Unit
ABSTRACT SUPERTYPE OF (
  ONEOF(entity_definition,entity_implementation,configuration));
END_ENTITY;

ENTITY unit_definition
 ABSTRACT SUPERTYPE OF (ONEOF(entity_definition,package_definition))
 SUBTYPE OF ( Library_Unit);
   unit_name : Meaningful_Name;
DERIVE
   owning_library : library :=
   USEDIN(Library_Unit,'library.has_unit_definitions')[1];
UNIQUE
   unit_name,owning_library;
END_ENTITY;

ENTITY entity_definition
SUPERTYPE OF ( simulatable_entity )
SUBTYPE OF ( unit_definition, template, Instantiable_Unit );
   has_implementations : LIST [0 : ?] OF entity_implementation;

— private definitions used by the design unit to ...
— define types, variables, signals used by the checks or the
— implementations
— private definitions  of (for example) the contents of a ROM
  has_types      :   OPTIONAL SET [1 : ?] OF VHDL_TYPE;
  has_signals    :   OPTIONAL SET [1 : ?] OF signal;
  local_values   :   OPTIONAL SET [1 : ?] OF Undefined;
— private statements used by the design unit to
— check it is being used within its design constraints.
— They will involve timing and modelling changes in signal values.
```

```
  has_checks : OPTIONAL SET [1 : ?] OF passive_process;
END_ENTITY;

ENTITY template
 ABSTRACT SUPERTYPE OF
   (ONEOF(entity_definition,component_template,block));
— Interface of template to the world
 has_ports :   OPTIONAL SET [1 : ?] OF port;
— parameters to be passed to the template
 has_generics : OPTIONAL SET [1 : ?] OF generic_parameter;
END_ENTITY;

ENTITY simulatable_entity SUBTYPE OF (entity_definition );
WHERE
fully_configured : TRUE;
END_ENTITY;

ENTITY implementation_base
  ABSTRACT SUPERTYPE OF (
   ONEOF(behavioural_implementation,structural_implementation)
   AND ONEOF(block, entity_implementation));
  has_actions : OPTIONAL SET [1 : ?] OF action;
  has_parameterised_structure : OPTIONAL SET [1 : ?] OF for_generate;
  has_conditional_structure : OPTIONAL SET [1 : ?] OF if_generate;
  has_embedded_structure : SET [0 : ?] OF block;
  has_signals  : SET [0 : ?] OF signal;
END_ENTITY;
ENTITY structural_implementation
SUPERTYPE OF (mixed_implementation)
 SUBTYPE OF(implementation_base);
  has_templates : OPTIONAL SET [1 : ?] OF component_template;
— The binding between components and sub–entities may be performed here
  has_configurations :  OPTIONAL SET [1 : ?] OF instance_configuration;
  has_structure : SET [1 : ?] OF component_instantiation;
END_ENTITY;
ENTITY mixed_implementation
SUBTYPE OF (behavioural_implementation,
   structural_implementation);
END_ENTITY;
ENTITY entity_implementation
SUBTYPE OF (  Library_Unit, implementation_base ,  Instantiable_Unit );
  is_implementation_of : entity_definition;
END_ENTITY;
```

```
— The configuration definition
ENTITY configuration
 SUBTYPE OF ( Library_Unit,  Instantiable_Unit );
   configuration_name : Meaningful_Name;
   design : entity_definition;
   sub_configurations : SET [0 : ?] OF instance_configuration;
WHERE
configuration_rules: TRUE;
— if architecture containing component has configuration use it
— otherwise it must be provided by this configuration
— duplicate configurations are not allowed
— should consider default binding indications (5.2.2)
— Also if an architecture is specified then it need not exist (yet)
— unless this is elaboration
END_ENTITY;

ENTITY package_definition SUBTYPE OF ( unit_definition );
— data types supported by the simulator/cad system
— data types used in the design...
 has_templates : OPTIONAL SET [1 : ?] OF component_template;
— i/o support
END_ENTITY;

— The package body may not be necessary
ENTITY package_implementation SUBTYPE OF ( Library_Unit );
  is_implementation_of :  package_definition;
END_ENTITY;

END_SCHEMA;

SCHEMA allowable_types;
— This schema defines the permissible VHDL types.
— It has been left empty in this abbreviated model
END_SCHEMA;


SCHEMA structural_vhdl;
REFERENCE FROM vhdl(template, implementation_base, entity_definition);
REFERENCE FROM vhdl(Library_Unit, entity_implementation,
 simulatable_entity);
REFERENCE FROM vhdl(Meaningful_Name, Undefined);
REFERENCE FROM allowable_types(VHDL_TYPE,VHDL_VALUE,
```

```
  allowable_signal_type);

ENTITY port
 SUPERTYPE OF (ONEOF(input_port, driving_port));
  default_value : VHDL_VALUE;
  has_value : OPTIONAL VHDL_VALUE;
END_ENTITY;

ENTITY input_port SUPERTYPE OF (inout_port) SUBTYPE OF (port) ;
END_ENTITY;
ENTITY driving_port
 SUPERTYPE OF (ONEOF(output_port, inout_port, buffer_port))
 SUBTYPE OF (port);
END_ENTITY;
ENTITY output_port SUBTYPE OF (driving_port);
END_ENTITY;
ENTITY inout_port SUBTYPE OF ( input_port, driving_port);
END_ENTITY;
ENTITY buffer_port SUBTYPE OF (driving_port);
END_ENTITY;

ENTITY block SUBTYPE OF ( template, implementation_base );
   port_associations : OPTIONAL SET [0 : ?] OF port_instantiation;
   generic_associations : OPTIONAL SET [0 : ?] OF generic_implementation;
END_ENTITY;

ENTITY for_generate;
  sub_instance_list : SET [0:?] OF component_instantiation;
  sub_loop_generate_list : SET [0:?] OF for_generate;
  sub_conditional_list : SET [0:?] OF if_generate;
  sub_block_list : SET [0:?] OF block;
END_ENTITY;

ENTITY if_generate;
  sub_instance_list : SET [0:?] OF component_instantiation;
  sub_loop_generate_list : SET [0:?] OF for_generate;
  sub_conditional_list : SET [0:?] OF if_generate;
  sub_block_list : SET [0:?] OF block;
 DERIVE
  condition : BOOLEAN := calculate_condition;
  — This is really a function of entities in the environment of the generate
  — and probably needs to wait on a new release of EXPRESS
 WHERE
```

```
Nothing_produced_if_condition_false: TRUE;
— If the boolean condition if FALSE then the generation is not performed
END_ENTITY;

ENTITY generic_parameter;
   type_of_parameter : VHDL_TYPE;
   default_value     : VHDL_VALUE;
WHERE
value_is_of_type : TRUE;
END_ENTITY;

ENTITY signal
SUPERTYPE OF (ONEOF(signal_parameter,signal_instance));
 is_type   : allowable_signal_type;
 has_value : VHDL_VALUE;
WHERE
— has_value is of type is_type
correct_type : TRUE;
END_ENTITY;

ENTITY signal_instance
SUBTYPE OF (signal);
 history   : LIST [0 : ?] of VHDL_VALUE;
WHERE
correct_type: TRUE;
— has_value and history elements are of type is_type
END_ENTITY;

ENTITY signal_parameter
SUBTYPE OF (signal);
END_ENTITY;

ENTITY component_template SUBTYPE OF (template);
— Optional binding to an entity_definition for open configurations
 has_actual : OPTIONAL  entity_definition;
— The name is significant if there are no configurations for the component
 component_name : Meaningful_Name;
WHERE
actual_is_not_a_configuration: TRUE;
END_ENTITY;

ENTITY component_instantiation;
 component_definition :  template;
```

```
— It will be matched with the entity_definition when a configuration statement
— for the component is added to the database
  port_associations : OPTIONAL SET [0 : ?] OF port_instantiation;
— This IS the generic, not a pointer to anywhere, though other generics
— may be used by the actual configuration
  generic_associations : OPTIONAL SET [0 : ?] OF generic_implementation;
END_ENTITY;

ENTITY port_instantiation
SUPERTYPE OF (ONEOF(driving_port_instantiation,
              driven_port_instantiation));
— This entity is (as in the EDIF model) not a real CAD object but
— establishes a relationship between the port in the template/entity
— and the signal driving that port
  this_port : port;
  actual : signal;
  actual_conversion : OPTIONAL Undefined;
  — function to convert from type of actual
  formal_conversion : OPTIONAL Undefined;
  — function to convert to type of formal
WHERE
— some relationship between the types (formal port => actual port)
— Supplying of required parameters
correct_for_mode: TRUE;
same_value: this_port.has_value = actual.has_value;
END_ENTITY;

ENTITY driven_port_instantiation SUBTYPE OF (port_instantiation);
 SELF\port_instantiation.this_port : input_port;
 has_drivers    :    SET [0 : ?] OF signal;
DERIVE
— actual signal value from has_resolution and the actual vector of inputs
  value_of_port : VHDL_VALUE :=
                calculate_resolution(has_resolution,has_drivers);
WHERE
Resolution_function: check_resolution(has_resolution,has_drivers);
END_ENTITY;

RULE check_fanin FOR (signal);
WHERE
xxx: NOT ISEMPTY(QUERY
       (fanin <* USEDIN(signal,'driving_port_instantiation.actual') |
     EXISTS(fanin.has_resolution)));
```

```
END_RULE;

ENTITY driving_port_instantiation SUBTYPE OF (port_instantiation);
  SELF\port_instantiation.this_port : driving_port;
END_ENTITY;

ENTITY generic_implementation;
  formal : generic_parameter;
  actual : VHDL_VALUE; — elaboration time value of constant/variable
END_ENTITY;

ENTITY open_configuration SUPERTYPE OF (instance_configuration);
END_ENTITY;

ENTITY instance_configuration
SUPERTYPE OF (ONEOF(full_configuration,entity_configuration))
SUBTYPE OF (open_configuration);
   object_instance: SET [0 : ?] OF component_instantiation;
   — for all:cmp... where cmp is never instantiated would produce an
   — empty list
— This may be an entity/architecture/configuration or a deferral
   actual : Library_Unit;
   port_associations : OPTIONAL SET [0 : ?] OF port_instantiation;
— This IS the generic
   generic_associations : OPTIONAL SET [0 : ?] OF generic_implementation;
— This has the effect (in elaboration – not modelled) of replacing
— the component definition with a Library_Unit
WHERE
only_once_per_component: TRUE;
port_mapping: TRUE;
generic_mapping : TRUE;
actual_entity_is_simulatable: TRUE;
— this last constraint applies to elaboration time
END_ENTITY;

ENTITY full_configuration SUBTYPE OF (instance_configuration);
   SELF\instance_configuration.actual : entity_implementation;
END_ENTITY;
ENTITY entity_configuration SUBTYPE OF (instance_configuration);
   SELF\instance_configuration.actual : simulatable_entity;
END_ENTITY;

FUNCTION check_resolution(
```

```
    drivers : AGGREGATE OF signal) : BOOLEAN;
END_FUNCTION;

FUNCTION calculate_resolution(drivers : AGGREGATE OF signal)
                        : VHDL_VALUE;
END_FUNCTION;

FUNCTION calculate_condition : BOOLEAN;
END_FUNCTION;

FUNCTION is_empty(agg : GENERIC) : BOOLEAN;
END_FUNCTION;

END_SCHEMA;
```

Formal verification of VHDL descriptions in Boyer-Moore : first results

Dominique BORRIONE*, Laurence PIERRE**, Ashraf SALEM*

*IMAG/ARTEMIS, B.P. 53X, 38041 Grenoble Cedex, France
tel: +33.76.51.46.04; fax: +33.76.51.96.37

**Université de Provence, 3 Place Victor Hugo, 13331 Marseille Cedex 3, France
tel: +33.91.10.61.10; fax: +33.91.10.61.02

1. Introduction

In order to achieve bug-free designs, an important first step is to ascertain their functional correctness. Except for very small circuits, the functional verification of a design by exhaustive simulation is impractical. This is why, for the last ten years, considerable research efforts have gone into finding theoretical models, proof methods and efficient algorithms to perform the *formal verification* of a design correctness. Formally verifying a circuit design consists into *proving* that, for all acceptable initial state values and for all possible input values, the design *implementation* (how it is built) realizes its *specification* (its expected behavior). Results obtained in the last years show that this approach is now applicable to reasonnably complex circuits.

To gain acceptance in the community of hardware designers, formal verification should be embedded, as one of the available design tools, into a complete CAD system [BP88]. Thus, the proof subsystem should take as input the same user interface descriptions which are used for the other design tasks. The standardization of VHDL, and the considerable interest raised in the community by VHDL-based software, lead us to apply these principles to the formal verification of VHDL descriptions, which involves defining the VHDL semantics in terms of a formally

J. Mermet (ed.), VHDL for Simulation, Synthesis and Formal Proofs of Hardware, 227–243.

manipulable model (a functional model in our case [Pa86]), and writing a translator from VHDL to the input format of provers for this model.

Previous work reported the principles of a first prototype prover for a subset of VHDL limited to bit level descriptions [SB90]. In parallel, the efficiency of the Boyer-Moore theorem prover [BM88] in formally verifying more abstract hardware descriptions has been shown [Hu86, BT89, Pi89a, VC90]: its inductive inference mechanisms make it possible to prove automatically repetitive structures and certain types of sequential circuits. The present paper builds on these works. We defined a method by which VHDL descriptions are translated in terms of the Boyer-Moore logic, and the appropriate theorems are generated, with minimal additional informations supplied by the designer. This will allow us to provide an automatic proof system for a larger subset of VHDL primitives.

2. Principles of the formal verification

A hardware unit is described in VHDL [IE88] using a design entity . The design entity consists of two parts: the entity declaration and the architecture body. Each design unit has only one entity declaration, which essentially describes its parameters and input-output ports; but it may have several architecture bodies, describing alternative implementation hypothesis, or different description levels. Our method [SB90] to formally verify a circuit described in VHDL is based on a functional approach. Starting from two architecture bodies of the same design entity (one is considered as "specification", and the other one is the "implementation"), the methodology consists in :

1) translating the VHDL descriptions of the two architecture bodies into LISP-like expressions which give a functional representation of either their structures or their behaviours;
2) proving, for each output port, the equivalence between their two corresponding functions.

The Boyer-Moore theorem prover is based on the principles of recursion and induction. Therefore, it can be applied with satisfactory results to categories of circuits that can naturally be expressed by recursive functions; it includes most of sequential circuits, and regular repetitive (combinational or sequential) architectures.

- For the sequential circuits, the recursion implicitely represents the transformation of the circuit state between two time steps,
- for repetitive devices, the recursive pattern translates the regularity of the structure.

3. Proof of sequential circuits

Following a reasoning similar to the one of W.Hunt [Hu86], we translate each VHDL architecture into a recursive function representing the whole system (this is the function denoted F_{seq} below); state variables are set as parameters of this function and updated in each recursive call. The recursive form of this global function describes the transformation from the current state of the circuit into its next state. This function must have an "acceptable" recursive formulation : the "measure" of at least one parameter (which usually represents time) must decrease in the recursive calls according to a "well-founded" relation.

State variables are primary inputs and "irreducible" internal variables (usually the memory elements). After a first translation step, a non-recursive expression is associated with each primary output and memory element; it is automatically generated from the VHDL description of the circuit, according to specialized translation principles and an internal variables elimination algorithm. This set of functional forms is used in the definition of the function F_{seq}.

3.1 General form

Let us consider (at the RT Level) a synchronous sequential circuit, with :

- the vector of primary inputs $\mathfrak{I} = (i_1, i_2, \ldots, i_n)$,
- the vector of primary outputs $\mathfrak{O} = (o_1, o_2, \ldots, o_m)$,
- and the vector of internal state variables $\mathfrak{S} = (s_1, s_2, \ldots, s_q)$

This device is assumed to be synchronized by a master clock, and we consider a discrete time scale which corresponds to the clock rising edges of this clock.The most general pattern of the recursive function that we associate with the whole circuit is :

$$\begin{aligned} F_{seq} = \lambda\,(I, S)\;.\; & \textit{if } \textit{length}(I) = 1 \\ & \textit{then } U_{\mathfrak{O}}(\textit{head}(I), S) \\ & \textit{else } F_{seq}(\textit{tail}(I), U_{\mathfrak{S}}(\textit{head}(I), S)) \end{aligned}$$

where
- I is the stream of the current and past values of the vector $\mathfrak{I}$. The function "head" extracts the first element of the stream, and the function "tail" extracts the rest of the stream,
- $U_{\mathfrak{O}}$ is the vectorial function which updates the vector of outputs $\mathfrak{O}$,
- and $U_{\mathfrak{S}}$ is the vectorial function which updates the vector of state variables $\mathfrak{S}$.

The recursive definition of F_{seq} models the evolution of the system between two successive time units, the vector of the state variables is updated according to the term $U_{\mathcal{S}}$ (head(I), S). This function gives as result the values of the outputs computed by means of the current values of the inputs and state variables, by the term $U_{\mathcal{O}}$ (head(I),S).

Moreover, the non-recursive function which is associated with the initial computation step is :

$$F_{seq}\text{-}init = \lambda I \,.\, F_{seq}(I, \mathcal{S}_0) \quad \text{where } \mathcal{S}_0 = (s_1(0), s_2(0), \ldots, s_q(0))$$

i.e. S is initialized with the values of the state variables at time 0. To describe the behaviour of this circuit during the first t time units, F_{seq}-*init* must be invoked by :

$$F_{seq}\text{-}init\,(<\mathcal{I}(0), \mathcal{I}(1), \mathcal{I}(2), \ldots, \mathcal{I}(t)>)$$

For the specification, which is expressed by another VHDL architecture of the same entity given at a different level of abstraction, the Boyer-Moore interpretation is also a tail-recursive function written at this level of abstraction. The theorem to be proven states the equivalence between these functional forms, assuming the appropriate typing hypotheses upon the variables.

3.2 Case of iterative arithmetic circuits

As far as iterative arithmetic circuits are concerned, the previous pattern can be simplified because this category of devices has the following characteristics :

- the inputs are positive integers,
- one of the outputs (whose corresponding function is denoted below U_O) holds the result of the computation and is meaningful only after a number of time steps equal to the initial value of one of the inputs, denoted i_T. Therefore, this input models the discrete time, and decreases in the recursive calls; the values of the other inputs are unchanged. The associated template no more needs streams.

The function F_{seq} is built according to the following simplified template :

$$\begin{aligned} F_{seq} = \lambda\, &(i_1, \ldots, i_T, \ldots, i_n, S)\,. \\ &\mathit{if}\ i_T = 0 \\ &\mathit{then}\ U_O(i_1, \ldots, i_T, \ldots, i_n, S) \\ &\mathit{else}\ F_{seq}(i_1, \ldots, i_T - 1, \ldots, i_n, U_{\mathcal{S}}(i_1, \ldots, i_T, \ldots, i_n, S)) \end{aligned}$$

For the specification, a simple non-recursive function is usually sufficient. The basic arithmetical operators are (recursively) predefined in the Boyer-Moore environment that we restore at the beginning of each proof session. Thus, the high-level VHDL description which expresses the specification only has to be translated into a simple function given in terms of these basic operators. In order to verify that the circuit performs the expected computation, we have to prove a theorem which states the equivalence between these two functions; this property is verified under the hypothesis that the input variables belong to the type of natural numbers.

However, an important feature of the Boyer-Moore system (and of induction in general) is that it can prove general properties more easily than particular cases. The equality of the previous lemma is a particular case because it involves constants; thus it cannot be proved directly by induction. A more general theorem must be submitted to the Boyer-Moore environment; this theorem can be automatically generated according to general principles about generalization, explained in [Pi90] : the new equality is obtained by expanding N times the two parameters of the initial equality, where N is the minimum number of expansions of the definition of F_{seq} required for the accumulating output to be different from a natural constant or a "Skolem constant" (see e.g. [Mo75]).

3.3 Example: a multiplier by iterated additions

The above discussion is now illustrated on the example of a multiplier, taken from [Go84], which computes the product of two natural integers X and Y by iterated additions; its VHDL description is given on Figure 1, its design is displayed on Figure 2.

The specification of this device is that, after X time units, the boolean output DONE is set to "true", and the result of the computation X*Y appears on the output OUT. For the specification of the multiplication, we give a higher-level architecture description in terms of arithmetic operators. The "*" operator of VHDL will be translated into the predefined recursive operator "times" in Boyer-Moore. A partial high-level specification is given by the following VHDL description :

```
architecture spec of multiplier is
begin
   outp <= x * y;
end spec ;
```

It is easily translated into the following Boyer-Moore function :

```
(defn Mult-spec (x y)
   (times x y))
```

```
entity multiplier is
    port ( x, y : in integer; clkedge : in boolean;
           outp : out integer; done : out bit);
end multiplier;

architecture implementation of multiplier is
signal l1, l3, l4, l5, l6, l10, l11, l12, l2, outpt : integer;
signal l7, l8, l9, t :bit;
signal donet : bit := '1';
begin
   l1 <= x when donet = '1' else l3;        l3 <= 0 when l2 = 0 else l2 -1;
   l4 <= 0 when l3 = 0 else l3 -1;          l5 <= 0 when x = 0  else x - 1;
   l6 <= l5 when donet = '1' else l4;       l7 <= '1' when (l6 = 0) else '0' ;
   l8 <= '1' when (y = 0) else '0';         l9 <= '1' when  (x = 0) else '0';
   l10 <= 0 when l9 = '1' else y;           l11 <= l10 when donet = '1' else l12;
   l12 <= l10 + outpt;                      t <= l7 or l8;
   clkblock: block (clkedge)
   begin
       donet <= guarded t;
       l2 <= guarded l1;
       outpt <= guarded l11;
   end block ;
   done <= donet;
   outp <= outpt;
end;
```

Figure 1: VHDL description of the multiplier by iterated additions

Two Boyer-Moore functions "Mult" and "Mult-init" are automatically associated with the implementation of Figures 1 and 2. The first one is recursive and represents the evolution of the system during X time units; it is built according to the pattern of F_{seq} above, where X plays the part of the decreasing input. The second one describes the initial step.

In processing the description of the implementation, the elimination program finds that the circuit contains 3 memory elements, L2, DONE and OUT (in fact, DONE and OUT are also primary outputs). Its gives the functional forms associated with these "irreducible variables" :

```
(de l2 (done x l2)
  (if done                              ; l2 = if done
      x                                 ;         then x
      (if (equal l2 0) 0 (sub1 l2))))   ;         else (if l2=0 then 0 else l2-1)
```

```
(de out (done x y out)
   (if done                                 ; out = if done
       (if (equal x 0) 0 y)                 ;          then (if x=0 then 0 else y)
       (+ (if (equal x 0) 0 y) out)))       ;          else (if x=0 then 0 else y)+out

(de done (done x l2 y)
   (or (equal (if done
                  (if (equal x 0) 0 (sub1 x))
                  (if (equal (if (equal l2 0) 0 (sub1 l2)) 0)
                      0
                      (sub1 (if (equal l2 0) 0 (sub1 l2)))))
              0)
       (equal y 0)))
```

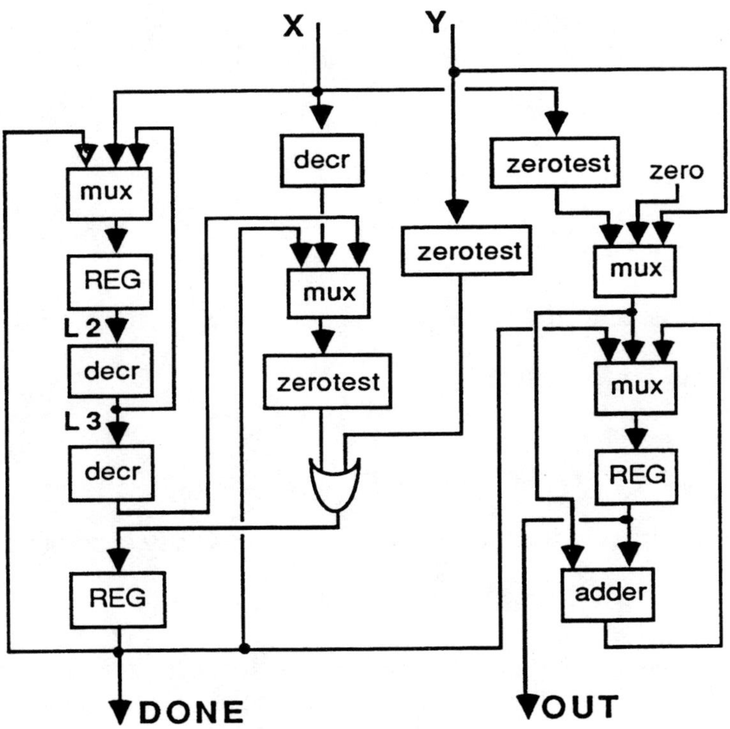

where : mux(c,e1,e2,s) ↔ s = if c then e1 else e2
decr(e,s) ↔ s = if (e=0) then 0 else e-1
zerotest(e,s) ↔ s = (e=0)
adder(e1,e2,s) ↔ s = e1+e2

Figure 2: Multiplier by iterated additions

The functions which represent the whole circuit are :

```
(defn Mult (x y done out l2)
   (if (zerop x)
      (if done out 0)  ; result given by "out" only if "done" holds `1 at the same time.
      (Mult (sub1 x)
            y
            (or (equal (if done          ; updating of "done"
                              (if (equal x 0) 0 (sub1 x))
                              (if (equal (if (equal l2 0) 0 (sub1 l2)) 0)
                                  0
                                  (sub1 (if (equal l2 0) 0 (sub1 l2)))))
                        0)
                 (equal y 0))
            (if done                     ; updating of "out"
               (if (equal x 0) 0 y)
               (plus (if (equal x 0) 0 y) out))
            (if done                     ; updating of "l2"
               x
               (if (equal l2 0) 0 (sub1 l2))))))
```

and :

```
(defn Mult-init (x y)        ; boolean constants, denoted '0' and '1' in the VHDL
   (Mult x y t 0 0))         ; syntax, correspond to "f" and "t" in Boyer-Moore.
```

The lemma associated with the verification of the specification is :

```
(prove-lemma Mult-ok ()
   (implies (and (numberp x) (numberp y))
            (equal (Mult-init x y)
                   (Mult-spec x y))))
```

But the equality involved in *Mult-ok* straightforwardly becomes :

```
(equal (Mult x y t 0 0) (times x y))
```

The induction mechanism of the Boyer-Moore system is unable to directly prove this lemma, essentially because the accumulating parameter "out" of "Mult" is assigned the constant value 0.

If we expand once the two terms of the equality, adding the appropriate hypotheses to avoid being in the terminal cases, we obtain :

```
(implies (and (numberp x) (numberp y) (not (equal x 0))
              (not (equal (sub1 x) 0)) (not (equal y 0)))
         (equal (Mult (sub1 x) y f y x)
                (plus y (times (sub1 x) y))))
```

Now, the accumulating parameter of "Mult" is assigned the value "y" which is not a natural constant, but is equal to one of the original "Skolem constants" x and y. For this reason, the above expression cannot be generalized, and we must expand once more the original statement; we get :

```
(implies (and (numberp x) (numberp y) (not (equal x 0))
              (not (equal (sub1 x) 0)) (not (equal y 0)))
              (not (equal (sub1 (sub1 x)) 0)))
         (equal (Mult (sub1 (sub1 x)) y f (plus y y) (sub1 x))
                (plus y (plus y (times (sub1 (sub1 x)) y))))))
```

which can be generalized, knowing that "sub1" give results in the domain of positive integers, into :

```
(implies (and (numberp y) (not (equal y 0))
              (numberp v) (not (equal v 0)) (not (equal (sub1 v) 0)))
         (equal (Mult (sub1 v) y f (plus y y) v)
                (plus y (plus y (times (sub1 v) y))))))
```

This formulation is satisfying because the accumulating parameter is neither equal to a natural constant, nor to a Skolem constant. Therefore, the following lemma has to be proven :

```
(prove-lemma Mult-ok-gen (rewrite)
  (implies (and (numberp y) (not (equal y 0))
                (numberp v) (not (equal v 0)) (not (equal (sub1 v) 0)))
           (equal (Mult (sub1 v) y f (plus y y) v)
                  (plus y (plus y (times (sub1 v) y))))))
  ((induct (times x y))))        ; for an induction on the parameter "v" instead of "y".
```

The CPU times for the proofs (on a Sun SPARCstation 2) are 3 seconds for *Mult-ok-gen*, and 1.9 seconds for *Mult-ok*.

4. Repetitive structures

This category includes the circuits which consist of N instances of a same module (using a *for...generate* component instantiation statement). The inductive properties of the Boyer-Moore prover allow to recursively define these devices and to realize the proof by induction on the structure. Description and proof are here general : we process N-bit circuits, where N is a positive integer *generic* .

The specification corresponds usually to a higher level function which takes as parameters natural numbers. Therefore, the statement to be proved calls on a function which converts bit-vectors into natural numbers, named "bv-to-nat".

We also need the "bit-vector" data type, it is defined by means of a "shell" (abstract data type) in the Boyer-Moore system [Hu86] :

```
(add-shell bitv                                   ; constructor.
           btm                                    ; bottom object.
           bitvp                                  ; recognizer.
           ((bit (one-of truep falsep) false)     ; first bit of the vector,
                                                  ; default value "false".
            (vec (one-of bitvp) btm)))            ; rest of the vector,
                                                  ; default value "btm".
```

Examples of combinational regular repetitive structures, including a N-bit ripple-carry adder [Hu86], a right to left or a left to right iterative comparison network, a multiplexer etc... [Pi89b] have been reported previously. In this paper, we present the translation and proof of the SN74163 presettable counter [TT76], as an example of sequential repetitive structure. A double recursion is involved in this example: a recursive function on time, portraying the sequential nature (a register) of the base cell, is called in a recursive function on the structure, associated with the overall architecture.

This N-bit counter presents a vectorial output, and a carry is propagated between the cells; Figure 3 shows the structure of this counter.The basic cell of this structure includes a JK-flipflop, described below :

```
entity JKFF is
   port (j,k : in bit; clkedge : in boolean; q : out bit);
end JKFF;

architecture impl of JKFF is
signal qt : bit;
begin
   clkblock : block (clkedge)
   begin
      qt <= guarded ((j and not qt) or (not k and qt));
   end block;
   q <= qt;
end impl;
```

In the Boyer-Moore formalism, the JK-flipflop is defined by means of a recursive function where the recursion represents the evolution of the device between two time steps :

```
(defn jkff (j k q time)
   (if (zerop time)
      q
      (jkff j k (or (and j (not q)) (and (not k) q)) (sub1 time))))
```

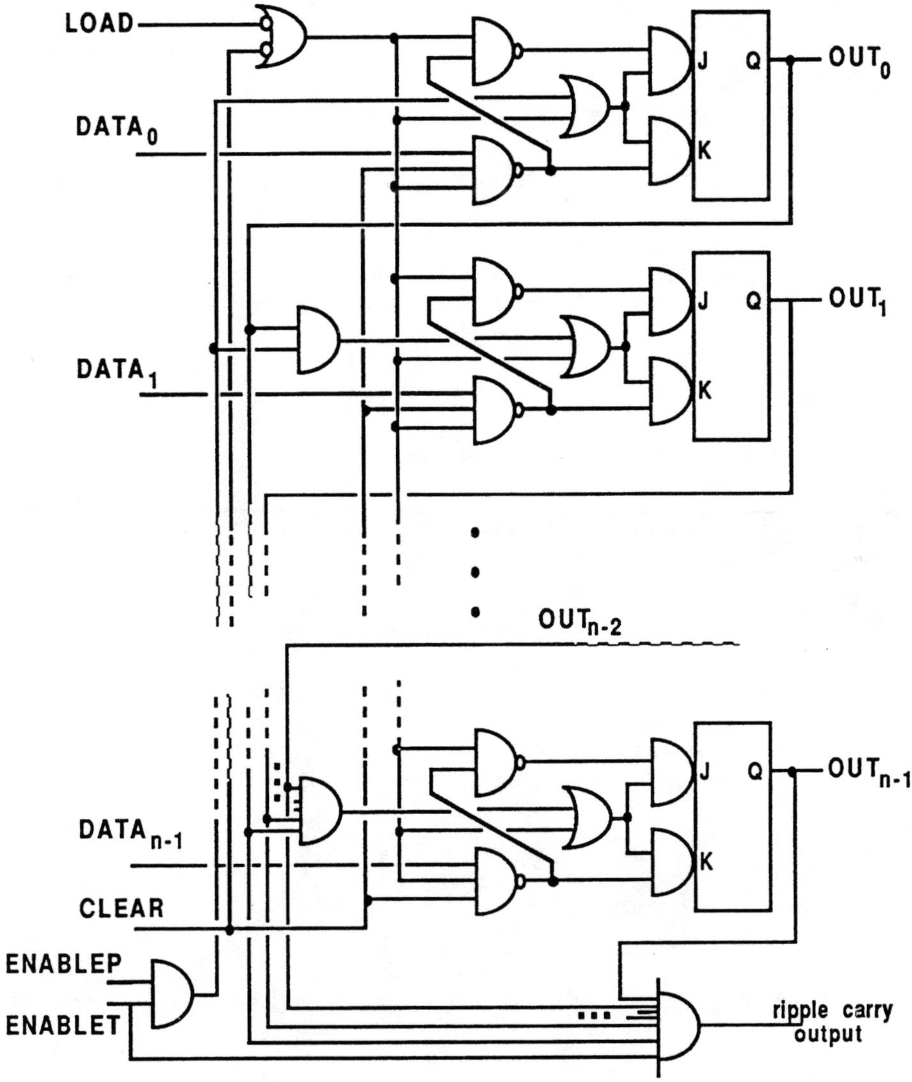

Figure 3 : N-bit SN74163 counter

The basic cell of the counter is given by :

```
entity COUNT_CELL is
    port ( data, cin, nclr, nload : in bit; clkedge : in boolean;
           outc, cout : out bit);
end COUNT_ CELL ;

architecture impl of COUNT_CELL is
component jkff_comp port (j, k : in bit; clkedge : in boolean; q : out bit);
end component;
```

```
signal inj, ink, outct : bit;
begin
    jkblock : block
    for jkff1:jkff_comp use entity JKFF(impl);
    begin
        jkff1: jkff_comp port map(inj,ink,clkedge,outct);
    end block;
    inj <= (not (data and nclr and (not nload  or not nclr)) nand
            (not nload or not nclr)) and (cin or (not nload or not nclr));
    ink <=not (data and nclr and (not nload or not nclr)) and
            (cin or (not nload or not nclr));
    cout <=  cin and outct;  outc <= outct;
end impl;
```

The function *count_cell* associated with the output "outc" of the basic cell calls on the function "jkff", where the JK-flipflop evolves during 1 time unit. The function *count_cell_carry* corresponds to the output "cout" of the basic entity.

```
(defn count_cell (data nclr nload cin outc)
   (jkff (and (not  (and (not (and data nclr (or (not nload) (not nclr))))
                          (or (not nload) (not nclr))))
              (or cin (or (not nload) (not nclr))))
         (and (not (and data nclr (or (not nload) (not nclr))))
              (or cin (or (not nload) (not nclr))))
         outc
         1))

(defn count_cell_carry (cin outc)
   (and cin outc))
```

The N-bit counter of Figure 3 can be described by the generic entity below :

```
entity count is
   generic (n :integer);
   port  ( clkedge : in boolean ; nclr, nload, enablep, enablet : in bit;
           data : in bit_vector (0 to  n-1 );
           outc : out bit_vector  (0 to  n-1 );  rco : out bit);
end  count;

architecture impl of count is
component  count_cell_comp
   port  ( data, cin, nclr, nload : in bit; clkedge : in boolean;
           outc, cout : out bit);
end component;
signal carry :bit_vector (0 to n);
```

```
begin
   cellblock :block
   for all : count_cell_comp use entity work.count_cell(impl);
   begin
       b1: for j in 0 to n-1 generate
           b2: count_cell_comp port map ( data(j), carry(j), nclr, nload,
                                          clkedge, outc(j), carry(j+1));
       end generate;
   end block;
   carry(0) <= enablep and enablet;
   rco <= carry(n) and enablet;
end impl;
```

To describe the regularity of this device, we define a recursive function; each recursive call allows to compute the next bit of the vectorial output (with the "count_cell" call), and the output bit-vector is built by the function "bitv" :

```
(defn count (data nload enablep enablet nclr outc carry n)
   (if (and (bitvp data) (bitvp outc) (numberp n)
            (equal (size data) n) (equal (size outc) n))
      (if (equal n 0)
         (btm)
         (bitv (count_cell (bit data) nclr nload carry (bit outc))
               (count (vec data) nload enablep enablet nclr (vec outc)
                      (count_cell_carry carry (bit outc)) (sub1 n))))
      (btm)))
```

The truth table which gives the specification of this device is [TT76] :

NCLR	NLOAD	ENABLET	ENABLEP	action
F	⊥	⊥	⊥	reset
T	F	⊥	⊥	load DATA -> OUT
T	T	T	T	increment
T	T	F	⊥	no change
T	T	⊥	F	no change

where the symbol "⊥" means "don't care".

This is translated into the following high-level VHDL description :

```
architecture spec of count is
function Int_val (a : bit_vector) return integer is
--- conversion bit-vector -> natural
function Bit_val (n: positive; a : natural) return bit_vector is
--- conversion natural -> bit-vector
```

```
signal outt : bit_vector (0 to n-1);
begin
    clkblock : block (clkedge)
    begin
        outt <= guarded  bit_val (n,0) when (nclr = '0') else
                         data when (nload = '0') else
                         bit_val (n,Int_val(outt) + 1) when
                                   (enablet = '1' and enablep = '1') else
                         outt ;
      end block;
    outc <= outt ;
end spec;
```

And this VHDL description yields the Boyer-Moore function *count-spec* below :

```
(defn count-spec (data nload enablep enablet nclr outc n)
   (if (not nclr)
      (nat-to-bv 0 n)
      (if (not nload)
         data
         (if (and enablep enablet)
            (if (all-onesp outc)          ; "all-onesp" is the predicate which checks
               (nat-to-bv 0 n)            ; if the bit-vector only includes 1's.
               (nat-to-bv (add1 (bv-to-nat outc)) n))
            outc)))))
```

Therefore, we should verify the following theorem which exactly translates the specification above :

```
(prove-lemma count-ok ()
   (implies (and (bitvp data) (bitvp outc) (numberp n)
                 (equal (size data) n) (equal (size outc) n))
            (equal (count data nload enablep enablet nclr outc
                          (and enablep enablet) n)
                   (count-spec data nload enablep enablet nclr outc n))))
```

We remark that the parameters *enablep* and *enablet* appear in the term *(and enablep enablet)*, and nowhere else, neither in the definition of "count" nor in the specification; thus the expression *(and enablep enablet)* is "irrelevant", according to the Boyer-Moore terminology, and must be generalized into the boolean *c* . The theorem is rewritten into :

```
(prove-lemma count-ok ()
   (implies (and (boolp c) (bitvp data) (bitvp outc) (numberp n)
                 (equal (size data) n) (equal (size outc) n))
```

```
(equal (count data nload enablep enablet nclr outc c n)
       (if (not nclr)
           (nat-to-bv 0 n)
           (if (not nload)
               data
               (if c
                   (if (all-onesp outc)
                       (nat-to-bv 0 n)
                       (nat-to-bv (add1 (bv-to-nat outc)) n))
                   outc))))))
```

which is proved in 4.1 s, on a Sun SPARCstation 2.

5. Conclusion

A partial prototype is being implemented, limited to well identified classes of circuits, and based on a constrained description style in VHDL. The tool is organized in two phases:

- The first one takes as input the intermediate forms of the VHDL entity and architectures, produced after syntactic and semantic verification by commercial software [CL90], and builds the corresponding "proof oriented" functional representations of the circuit. This phase can also be used in the context of the automatic proof of VHDL with other demonstrators [Sa89].

- The second phase is interactive. A query-answer dialogue obtains from the designer the characteristic properties of the circuit (sequential vs repetitive structure, master-clock name, initial values, ...) necessary to determine the circuit type, and generate the appropriate function definitions in Boyer-Moore syntax. Then the theorem that states the equivalence between the specification and the implementation architectures is built. If necessary, the unfolding and generalization transformations are performed, before submitting the theorem to the prover.

The method described in this paper is applicable to synchronous sequential circuits, and to zero-delay models of repetitive combinational circuits. Future works will include enlarging the set of acceptable VHDL primitives, in particular concerning the various delay types, in order to verify a wider range of circuits and abstraction levels. This will require building a more comprehensive library of predefined theorems in Boyer-Moore, in order to make this automatic proof system more usable in real size design tasks.

References

[BM88] R.S Boyer, J.S.Moore : "A computational logic handbook", Academic Press, Inc.,1988.

[BP88] D. Borrione, J.L. Paillet, L. Pierre: "Formal Verification of CASCADE Descriptions", Proc. IFIP WG10.2 Working Conference on "The fusion of hardware designs and verification", Glasgow, July 1988 (North Holland).

[BT89] A. Bronstein, C.L. Talcott : "Formal Verification of Pipelines based on String-Functional Semantics". In "Formal VLSI Correctness Verification", L.Claesen Ed., North Holland, 1990 ISBN 0444 88689 3.

[CL90] CLSI : "VHDL Tool Integration Platform (VTIP)", CAD Language Systems, Rockville, MD, USA, April 1990

[Go84] M. Gordon : "LCF-LSM". Technical Report n°41. University of Cambridge (UK). 1984.

[Hu86] W. A. Hunt: "FM8501: a verified micro-processor", Institute for Computing Science, Univ. of Texas, Austin, USA, Technical report N° 47, Feb. 1986

[IE88] IEEE. IEEE standard VHDL Language Reference Manual, IEEE, 1988.

[Mo75] J.S. Moore : "Introducing iteration into the Pure Lisp theorem prover". IEEE Transactions on Software Engeneering, Vol.SE-1, n°3. September 1975.

[Pa86] J.L Paillet. "A functional model for descriptions and specifications of digital devices", Proc. IFIP Int. Working conf. "From HDL descriptions to guaranteed correct circuit designs", Grenoble, North Holland, September 1986.

[Pi89a] L. Pierre : "The formal proof of sequential circuits described in CASCADE using the Boyer-Moore theorem prover". In "Formal VLSI Correctness Verification", L.Claesen Ed., North Holland, 1990 ISBN 0444 88689 3.

[Pi89b] L. Pierre : "The Formal Proof of the "Min-max" sequential benchmark described in CASCADE using the Boyer-Moore Theorem Prover." In "Formal VLSI Correctness Verification", L.Claesen Ed., North Holland, 1990 ISBN 0444 88689 3.

[Pi90] L. Pierre : "Automatically generalizing theorem statements for the inductive proofs of some digital circuits". Rapport de recherche MAIUP n°90-02. Université de Provence (Marseille). March 1990.

[Sa89] A. Salem: "Formal verification of the JANUS on-line multiplier", R.R. N° 785-I, ARTEMIS, Grenoble, July 1989.

[SB90] A. Salem, D. Borrione: "Automatic formal verification of VHDL descriptions: a first prototype", R.R. N° 823-I ARTEMIS, Grenoble, France, June 1990.

[TT76] TTL Databook for design engineers, 2nd edition, Texas Instruments, INC. 1976.

[VC90] D. Verkest, L. Claesen, H. DeMan : "Correctness proofs of parameterized hardware modules in the Cathedral-II synthesis environment". Proc. European Design Automation Conference. Glasgow (U.K.). 12-15 March 1990.

Developing a Formal Semantic Definition of VHDL

PHILIP A. WILSEY
Computer Architecture Design Laboratory
Dept. of Electrical and Computer Engineering
University of Cincinnati, Cincinnati, OH 45221-0030
phil.wilsey@uc.edu (513) 556-4779

Abstract.
The work described in this paper is concerned with developing a formal semantic model of machine behavior that is compatible with the hardware description language VHDL. Currently, we have developed a formal model sufficiently robust to characterize many key components of VHDL. In particular, we have defined the propagation of signal values up and down the component hierarchy (complete with type conversion functions), bus resolution of signal values, process statements including the wait statement, transaction list definition, and the computation of driving values. The one main concept of VHDL yet to be included in this model is the delta delay.

1. Introduction

Hardware description languages have been in existence for approximately three decades. Some of the most well known HDLs are DDL [8], ISPS [22], and PMS [22]. However, the use of such languages has been restricted primarily to only documentation and simulation of computer systems. Recent investigations have been more interested in developing HDLs that can be used throughout the computer system design process, from design conception to hardware, and ultimately, software synthesis. Examples of languages that fall into this category are AADL [6], CONLAN [20], SBL [12], ACE [25], and VHDL [17; 18].

Naturally, as experience is gained with HDLs, new and more powerful languages capable of machine description across a wide range of abstraction levels are emerging. Hence, such languages are enjoying a more active use in the computer system design process. Unfortunately, with very few exceptions, these language usually lack a formal semantic definition and thus, even though an architecture is formally described using an HDL, the resulting description may still be subject to different, possibly conflicting, interpretations.

Although some attempts have been made toward formal definitions (*e.g.*, [4; 5; 9; 10; 11; 16; 19; 21; 24; 26]), the lack of a formal semantic definition for these languages is not an oversight on the part of the designer of the HDL. The real problem lies with the fact that no readily acceptable mechanism for specifying the formal semantics of temporal based languages really exists. Unfortunately, the traditional techniques for semantic definition of programming languages (*e.g.*, denotational semantics [13; 23], axiomatic semantics [15; 14], operational semantics, and so on) are, for one reason or another, inadequate for use in formally defining these languages.

This paper presents an approach for defining the semantics of VHDL using an interval temporal logic. The specific temporal logic to be used was originally developed by J. Allen [1; 2; 3] for reasoning about action and time. Thus far,

J. Mermet (ed.), VHDL for Simulation, Synthesis and Formal Proofs of Hardware, 245–256.

Interval Relation	*Pictoral Example*
`t before s`	`ttt sss`
`t equals s`	`ttt` `sss`
`t meets s`	`tttsss`
`t overlaps s`	`tttttt` `   ssssss`
`t during s`	`  ttt` `sssssss`
`t starts s`	`ttt` `ssssss`
`t finishes s`	`   ttt` `ssssss`

Fig. 1. Relationships of Time Intervals

we have successfully defined the semantics of a subset of the VHDL hardware description language.

The remainder of this paper is organized as follows. Section 2 introduces the interval temporal logic that we use for the formal definition. Section 3 presents the temporal elements, object declarations, and component instantiations. Section 4 provides the formal axioms specifying the definition of transaction lists, drivers, and driving values. Finally, Section 5 contains some concluding remarks.

2. An Interval Temporal Logic

In this work, we will use an interval temporal logic that was originally developed by Allen for reasoning about action and time [1; 2; 3]. Our main reason for selecting Allen's formalism is that it allows us to reason about the relationships of time intervals to one another as well as the relationships of actions to time intervals.

This logic is based upon time intervals rather than time points. This approach is taken so that one can easily reason about the occurrence of an action without concern as to the precise point in time that the action actually occurs. With respect to time intervals, Allen has identified seven relationships, with inverses, that can hold between two time intervals. These relationships are summarized in Figure 1.

Each of the interval relations of Figure 1 has a corresponding predicate in the logic. Let t and s each represent some time interval; then these predicates are:

EQUALS(t,s): t and s represent the same time interval,

MEETS(t,s): t is before s and there is not an interval r that is between t and s,

DURING(t,s): t is fully contained within s,

OVERLAPS(t,s): t begins before s begins, s begins before t finishes, and t finishes before s finishes,

STARTS(t,s): t and s begin together, but t finishes before s finishes,

FINISHES(t,s): t starts after s starts, but they finish together, and

BEFORE(t,s): t finishes before s begins and there is a time interval r that is between t and s.

Allen has defined a set of axioms for the behavior of these predicates. Among these are axioms asserting that each relationship is mutually exclusive of the others; others define the transitivity of the predicates. A detailed discussion of these can be found in [2]. For our purposes, we are concerned only with the fact that for any given interval, t, there exists another interval for each predicate of the logic such that the predicate holds. That is, given an interval t, there is an interval that meets t, that equals t, that starts with t, and so on.

Three additional predicates that will be useful in this work are:

$$\mathrm{IN}(t,s) \Leftrightarrow \mathrm{DURING}(t,s) \vee \mathrm{EQUALS}(t,s),$$
$$\mathrm{BEGINS}(t,s) \Leftrightarrow \mathrm{STARTS}(t,s) \vee \mathrm{STARTS}(s,t) \vee \mathrm{EQUALS}(s,t), \text{ and}$$
$$\mathrm{ENDS}(t,s) \Leftrightarrow \mathrm{FINISHES}(t,s) \vee \mathrm{FINISHES}(s,t) \vee \mathrm{EQUALS}(s,t).$$

Informally, the predicate IN states that either t and s represent the same interval or t is fully contained within s. The predicate BEGINS states that t and s begin together. The predicate ENDS states that t and s finish together.

Two other components of Allen's logic relate to *properties* and *events*. A *property* is a time dependent predicate defined on some aspect of the universe under consideration, or a particular state of that universe (*e.g.*, a static value of a program variable). An *event* is something that may cause a state change or the value of a particular predicate to change (*e.g.*, incrementing the value of a program variable). Corresponding to these two notions, Allen defines two predicates: HOLDS and OCCUR.

The predicate HOLDS takes a property, p, and a time interval, t, and is true if the predicate p is true throughout t. One of the most important properties of the HOLDS predicate is that if it is true for some predicate p and time interval t then it is also true for every time interval s contained in t. Formally, we state:

$$\mathrm{HOLDS}(p,t) \Leftrightarrow \forall s : (\mathrm{DURING}(s,t) \vee \mathrm{EQUALS}(s,t)) \Rightarrow \mathrm{HOLDS}(p,s).$$

The predicate OCCUR takes an event, e, and a time interval, t, and is true if the event e occurs in t and there is no subinterval s of t in which e occurs. Thus, the predicate OCCUR has the property that if it is true for an event e in time t, then it is not true for any subinterval of t. Formally, this property of the OCCUR predicate is defined by the axiom:

$$\mathrm{OCCUR}(e,t) \wedge \mathrm{DURING}(s,t) \Rightarrow \neg\, \mathrm{OCCUR}(e, s).$$

3. Elements of A Formal Semantic Model of VHDL

In the following subsections, we briefly demonstrate how Allen's logic can be used in the formal semantic definition of VHDL. More precisely, we develop a formal model of machine behavior that corresponds to VHDL. The intent of this work is to develop a formal machine model onto which any VHDL description can be mapped. For example, while the model has no notion of VHDL generate statements, it does support component instantiations. Thus, after elaboration, a VHDL description would readily map onto this machine model.

Throughout the remainder of this paper, a number of simplifying assumptions regarding the semantics of VHDL are made. More precisely:

1. Only signals of kind **register** are considered.
2. Delta delays are not considered. All signal assignment statements are assumed to have a non-zero delay and each process contains only one signal assignment statement to a signal s.
3. Transport and inertial delays are not considered. It is assumed that each transactions posted to a transaction list by a signal assignment statement will become an event.
4. Only ports of mode IN, OUT, or INOUT are defined.

In many cases, these simplifications occur only because of space constraints. For example, we discuss only signals of kind **register**, however signals of kind **bus** can also be readily modeled. In other cases, these simplifications present more serious deficiencies in the model that must be addressed. For example, the lack of delta delays.

AN OVERVIEW OF THE MODEL.

The basic approach of this work is to define the transaction lists, driving values, and effective values as sets. Elements of these sets are 2-tuples of the form:

$$\langle v, t \rangle$$

where v denotes a value and t denotes a time interval. These sets can then be examined to determine the state of the described machine at various time intervals. Thus, in the context of a simulation of the described system, the model can be examined at discrete time intervals for obtaining a single valued state for each signal of the described machine. In contrast, one can also examine the state of the machine over a larger time interval and observe the transitions that occur to the machine state during that interval.

The remainder of this paper proceeds as follows. First a formal model of time is developed. Second, signals and ports are formally characterized. Third, the propagation of values up and down the component hierarchy is defined. Fourth,

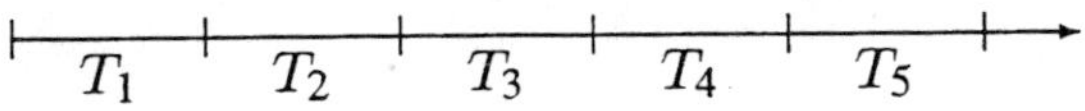

Fig. 2. Pictorial View of the Timekeeper

transaction lists, drivers, and signal driving values are defined. Finally, the process of interrogating the model to elicit a particular instance of the machine state is described.

Before presenting the formal definition, however, we must first introduce some notation. In this model, each element of VHDL is defined as a tuple. Due to the complexity and nesting of references through the tuples, we will always fully qualify references to the elements of tuples. For example, assume that we have the two definitions:

$$P = \{ \langle M, E, D, fn, T \rangle \}$$
$$pa = \langle F, fn_1, A, fn_2 \rangle$$

where the element F of pa is a member of P. The notation pa.F.M will then be used to reference the element M of F of pa. In addition, we will also use the function VAL(E,t) to indicate the value of the expression E in the time interval t.[1]

3.1. TIME.

In order to develop a technique for defining the semantic behavior of VHDL, we must first develop a unit time scale. This unit time scale will then denote time in the described machine. In this definition, the unit time scale is defined by an entity called the *timekeeper*. The timekeeper, *TK*, provides a single *universal temporal reference frame,* and is defined as:

$$\mathrm{TK} = \{\mathrm{tk(i)} : \mathrm{i} \geq 0 \wedge \mathrm{MEETS(tk(i),tk(i+1))}\}.$$

The timekeeper defines a set of unit time intervals that are adjacent to one another (Figure 2). Throughout the remainder of this paper, we will use, without qualification, the notation tk(i) to denote the ith element of the timekeeper.

VHDL has a predefined language environment that contains, among other things, the function NOW which returns the current simulation time. Formally, the function NOW is defined as:

$$\forall\, \mathrm{i} : \mathrm{i} \geq 0,\ \mathrm{VAL(NOW,tk(i))} = \mathrm{tk(i)}.$$

[1] Note that this VAL function must be used carefully. That is, it must be used only in restricted time intervals where the expression is constant.

3.2. SIGNALS.

The set of *signals* (S) in VHDL denote entities that hold values of the machine state. In this definition of VHDL, a distinction between signals declared with a signal declaration and port declarations is made. Port declarations will be discussed in the next section and they are not considered to be in the set S. Formally, a signal, $s \in S$, can be represented as a 6-tuple:

$$s = \langle E, D, fn, TY, K, I \rangle$$

where E is the set of *effective values*, D is the set of *driving values*, $fn{:}\{v\} \rightarrow v$ is the *resolution function* mapping a set of values into a single value, TY is the *data type* of the signal, K is the *signal kind* (`register` or `bus`), and I is the set of *initial values*.[2] In the case where no bus resolution function exists in the VHDL description of the signal, the function fn will be a function that simply returns the input value. In order to simplify exposition, the data type of a signal will not be further used or discussed in this paper and we will assume that all signals (ports and variables) have the same scalar type.

The resolution function fn is used by the simulation cycle to update the driving value of the signal from its drivers. The distinction between driving and effective values of a signal is not really necessary, but, its inclusion does contribute to the regularity of the model and therefore, the distinction is maintained.[3] Because we have included a distinction between driving and effective values, it is necessary to assert that the effective value always reflects the driving value. Formally, we state:

$$s.E = s.D$$

Distinctions between driving and effective values occur in the formals and actuals of the ports of an instantiated component. This is described in the next section.

3.3. COMPONENT INSTANTIATIONS AND PORT CONNECTIONS.

The set of *components* (C) in a VHDL description are entities that can be instantiated one or more times for the purposes of describing devices that process and transfer information of the machine state. Components are instantiated with the elements of its formal port clause associated with actual signals (or ports) of the instantiating environment. In the following two subsections, we provide a definition of a formal port clause and then we characterize the propagation of data up and down through the links of the component hierarchy.

[2] The initial values are provided in 2-tuple form, with the first element having a time component of tk(0) and all others with time components computed from tk(0). In the case where no explicit default value is given, an initial tuple of the form ⟨TY′LEFT,tk(0)⟩ is assumed.

[3] The VHDL language reference manual commonly refers to ports as signals. In our definition, we completely separate the set of signals from the set of port declarations. Thus, the statement that driving and effective values are the same *is* correct.

3.3.1. Ports.

The elements of the set of *formal ports* (P) represent all elements of the formal port clauses found in a VHDL description. Formally, a port declaration, $p \in P$, is represented as a 6-tuple:

$$p = \langle M, E, D, fn, TY \rangle$$

where M is the *mode* specifying direction for the flow of data through the port, E is the set of *effective values*, D is the set of *driving values*, $fn:\{v\} \rightarrow v$ is the *resolution function* mapping a set of values into a single value, and TY is the *data type*. In the case where no bus resolution function exists in the VHDL description of the port declaration, the function fn will be a function that simply returns the input value.

The definition of the set of driving values for ports of mode OUT or INOUT are determined by signal assignment statements as defined later in this paper. The effective values of ports of mode IN or INOUT are updated by the propagation of the effective values from the actual designator given in the component instantiation.

3.3.2. Port associations.

In conjunction with component instantiations there are the set of *port associations* (PA) that links the formal port names given in the component definition with the actual signal (or port) names of the instantiating environment. The set PA contains all of the port associations of a complete VHDL description. Formally, a port association, $pa \in PA$, can be represented as a 4-tuple:

$$pa = \langle F, fn_1, A, fn_2 \rangle$$

where $F \in P$ is the *formal* port designator, fn_1 is a type conversion function mapping values of the formal's type to values of the actual's type, $A \in S \cup P$ is the *actual* port designator, fn_2 is a type conversion function mapping values of the actual's type to values of the formal's type. In the case where no type conversion function(s) is specified in the VHDL description, fn_1 and/or fn_2 will be a function that simply returns the input value.

In the process of linking the names of the formal and actual parts, instantiating a component initiates the transfer of data up and down the component hierarchy. The data flows up the hierarchy by updating the driving values of the actuals from the driving values of the formals (for ports of mode OUT or INOUT only). Conversely, data flowing down the hierarchy causes the effective values of the formals to be updated with the effective values of the actuals. In either case, data flowing up or down, there may exist type conversion functions between the transmitted formal and actual values.

Formally, the transmission of effective values down the hierarchy (complete with type conversion) is expressed by the axiom:

```
        P1:   process begin
S1                  a <= a1;
S2                  wait on a until c1 for t1;
S3                  if c2 then
S4                         b <= b1;
S5                         wait for t2;
                    else
S6                         b <= b2;
                    end if;
S7                  d <= d1;
              end process;
```

Fig. 3. A VHDL Process Statement

$$\forall \text{ pa} : (\text{pa.F.M} = \text{IN}) \lor (\text{pa.F.M} = \text{INOUT}),$$
$$\langle \text{v}, \text{t} \rangle \in \text{pa.A.E} \Leftrightarrow \langle fn_2(\text{v}), \text{t} \rangle \in \text{pa.F.E}$$

Informally this axioms state that for all time intervals, the effective value of the formal will always be equal to the type converted value of the actual.

Similarly, data transmission up the hierarchy is expressed formally as:

$$\forall \text{ pa} : (\text{pa.F.M} = \text{OUT}) \lor (\text{pa.F.M} = \text{INOUT}),$$
$$\langle \text{v}, \text{t} \rangle \in \text{pa.F.D} \Leftrightarrow \langle fn_1(\text{v}), \text{t} \rangle \in \text{pa.A.D}$$

4. The Process Statement

Formally a process statement, PR, is defined as a 5-tuple:

$$PR = \langle\, D, SS, HEAD, SUCC, TR \,\rangle$$

where D is a set of *declarations*, SS is the set of *statements* contained in the process statement, $HEAD$ is a function to return the statement from SS that is initially evaluated at tk(0), $SUCC$ is a function of two arguments (a statement and a time interval) that determines which statement will be evaluated next, and TR is called the *transaction set*. All elements of this tuple except for TR, are amply illustrated in Figures 3 and 4. However, TR requires further discussion.

Elements of the transaction set, TR, are of the form:

$$TR = \langle\, D, T \,\rangle$$

$$
\begin{aligned}
&P1 = \langle D, SS, HEAD, SUCC \rangle \\
&\quad D = \emptyset \\
&\quad SS = \{S1, S2, S3, S4, S5, S6, S7\} \\
&\quad HEAD = S1 \\
&\quad SUCC(S1,t) = S2 \\
&\quad SUCC(S2,t) = S3 \\
&\quad SUCC(S3,t) = \begin{cases} S4, & \text{if } VAL(c2,t) = \text{true} \\ S6, & \text{otherwise} \end{cases} \\
&\quad SUCC(S4,t) = S5 \\
&\quad SUCC(S5,t) = S7 \\
&\quad SUCC(S6,t) = S7 \\
&\quad SUCC(S7,t) = S1
\end{aligned}
$$

Fig. 4. Formal Representation of a VHDL Process Statement

where D is a member of the set of signals and ports ($D \in S \cup P$) and T is the transaction list for that signal or port. Thus, for a process statement containing m signal assignment statements that assign values to n different destinations (signals), the set D will contain n elements; one for each distinct destination. A complete definition of the elements of T is shown below (Section 4.2). Before presenting its definition, however, we must first define the time intervals in which the statements of a process statement are evaluated.

4.1. PROCESS STATEMENT EVALUATION.

In an earlier paper, we have developed a set of definitions that determine the time intervals that each statement in a process statement is evaluated [26]. For each process statement, pr, [26] defines EV_{pr} as a set whose members define when each statement is evaluated. More precisely, an element of EV_{pr} is formed as a 2 tuple, $\langle s,t \rangle$, where s is a statement and t is a time interval. Each member of this set denotes that the statement s is evaluated in time t. In the remainder of this paper, this set will be referenced without further qualification. Before proceeding, however, it is necessary to review the formal representation of a signal assignment statement as presented in [26].

Formally, a signal assignment statement, sa, is represented as a 4-tuple:

$$sa = \langle D, TI, E, d \rangle$$

where D is the *destination* (signal or port name) of the statement, TI is the transport/inertial delay specifier,[4] E is the *expression* defining the new value for the signal, and d is the *delay* time value specified in the after clause. Since delta delays are being ignored, the delay, d, will always be nonzero.

[4] Although we have included this element of the tuple definition, we will hereafter ignore concerns for transport or inertial delays and instead assume that if a signal assignment statement schedules a transaction defining a new value of the destination, the new value *will* be assigned to the destination.

4.2. TRANSACTION SETS.

We must now select elements from the evaluation sets, EV_{pr} to determine the members of the transaction set for each process statement. Let PR be the set of all process statements in a VHDL description, then, for each pr ∈ PR, we write:

$$\begin{aligned}\forall\, \text{pr}, \langle \text{D,T}\rangle, \langle \text{s,t}\rangle : \text{pr} \in \text{PR} \wedge \langle \text{D,T}\rangle \in \text{pr.TR} \wedge \langle \text{s,t}\rangle \in EV_{pr} \wedge \text{s.D} = \text{D},\\ \text{T} = \{\langle \text{v,t}'\rangle : \langle \text{v,t}'\rangle \in \text{D.I} \vee\\ (\text{IN}(\text{t}',\text{t+s.d}) \wedge \text{v} = \text{VAL}(\text{s.E,t}))\}.\end{aligned}$$

Which simply states that the transaction set of each signal D is defined by either the initial values of the signal declaration (D.I) or by the evaluation of signal assignment statements.

4.3. SIGNAL DRIVING VALUES.

Finally, the driving values computed from the driving values of each transaction list must be defined. In this process, we will first define three elements for use, namely:

1. DRIVER(T,t): which defines the driver for transaction list T at time t,
2. ALL_DRIVERS(s,t): which defines the set of all drivers for signal s at time t, and
3. EVENTS(s): which defines the set of time intervals in which events occur to signal s.

Formally, DRIVER is defined as:

$$\begin{aligned}\text{DRIVER(T,t)} \equiv \langle \text{v,t}'\rangle : \langle \text{v,t}'\rangle \in \text{T} \wedge\\ (\text{EQUALS}(\text{t}',\text{t}) \vee \text{MEETS}(\text{t}',\text{t}) \vee \text{BEFORE}(\text{t}',\text{t})) \wedge\\ (\forall\langle \text{v}',\text{t}''\rangle : \langle \text{v}',\text{t}''\rangle \in \text{T} - \{\langle \text{v,t}'\rangle\},\\ \text{MEETS}(\text{t}'',\text{t}') \vee \text{BEFORE}(\text{t}'',\text{t}') \vee\\ \text{MEETS}(\text{t,t}'') \vee \text{BEFORE}(\text{t,t}''))\end{aligned}$$

which merely selects the member, ⟨v,t′⟩, from T such that it is prior to t and no other member of T has a timestamp between t′ and t.

In order to define a driving value for a signal, it is necessary to select all drivers for a signal during each time that an event for the signal occurs. Thus, we define the set ALL_DRIVERS as:

$$\begin{aligned}\text{ALL\_DRIVERS(s,t)} \equiv \{\, \langle \text{v,t}'\rangle : \langle \text{v,t}'\rangle = \text{DRIVER(T,t)} \wedge\\ \langle \text{s,T}\rangle \in \cup_{pr\in PR}\, \text{pr.TR}\,\}\end{aligned}$$

Finally, the times that an event occurs to signal s is defined by EVENT as:

$$\text{EVENT(s)} \equiv \{\, \text{t} : \exists\, \langle \text{s,T}\rangle \in \cup_{pr\in PR}\, \text{pr.TR} \wedge \langle \text{v,t}\rangle \in \text{T}\,\}.$$

Note that from this definition, it is quite simple to define the implicit signal s'EVENT; it is simply true in every time interval of the set EVENT(s).

Finally, defining the set of driving values for a signal is now defined as:

$$\forall s : s \in S, s.D = \{ \langle v,t \rangle : t \in \text{EVENTS}(s) \wedge v = s.\text{fn}(\text{ALL\_DRIVERS}(s,t)) \}$$

Informally this axiom defines a set of value/time pairs constituting the driving values for each signal s. Members of this set are defined to correspond with each event generated in all the transaction lists for that signal and furthermore, the value of each member is defined using the resolution function for the signal.

5. Conclusions

In this paper, we have presented an approach for developing a formal semantic model of VHDL. Probably the most difficult aspect of developing this formal definition has been in understanding the semantics of VHDL. Clearly the formal semantic definition presented in this paper is unambiguous and easy to follow. In addition to providing a clean, unambiguous semantic definition for users and tool builders alike, this definition can facilitate machine verification using techniques such as those developed by [7].

Finally, there remain a number of non-trivial extensions to this model that are necessary before it is sufficient for modeling the entire VHDL language set. In particular, (i) it will be necessary to refine the model of time to correctly represent delta delays, (ii) the inclusion of transport and inertial delays will also have to be developed for transaction list management, and (iii) signals of kind **bus** must be incorporated into the model.

Acknowledgements

This work was partially supported by the Defense Advanced Research Projects Agency under order no. 7056, monitored by the Federal Bureau of Investigation under contract number J-FBI-89-094. I would like to thank Praveen Chawla, Hal Carter, Paul Menchini, Dain Samples, Dave Barton, and Steve Levitan for their help with the semantics of VHDL or in the preparation of this manuscript.

References

1. ALLEN, J. F. An interval-based representation of temporal knowledge. *Proc. 7th Int. Joint Conf. on Artificial Intelligence* (Aug. 1981), 221–226.
2. ALLEN, J. F. Maintaining knowledge about temporal intervals. *Communications of the ACM 26* (Nov. 1983), 832–843.
3. ALLEN, J. F. Towards a general theory of action and time. *Artificial Intelligence 23* (1984), 123–154.
4. BORRIONE, D. The worker model of evaluation for compter hardware description languages. *Proc. 5th Int. Conf. on Computer Hardware Description Languages* (1981), 3–21.

5. BROCK, B. C., AND HUNT, JR., W. A. The formalization of a simple hardware description language. *Proc. Int. Workshop on Applied Formal Methods for Correct VLSI Design* (Nov. 1989), 83–98.
6. DAMM, W., DOEHMEN, G., MERKEL, K., AND SICHELSCHMIDT, M. The AADL/S* approach to firmware design verification. *IEEE Software 3*, 4 (July 1986), 27–37.
7. DECAMP, E. Methodology for the specification of temporal behaviours, applied to the formal verification of circuits. Tech. Rep. 580, Laboratoire ARTEMIS IMAG, France, 1986.
8. DULEY, J. R., AND DIETMEYER, D. L. A digital system design language (DDL). *IEEE Trans. on Computer C–17*, 9 (Sept. 1968), 850–861.
9. FILLIPPENKO, I. V. VHDL verification in the state delta verification system (SDVS). In *1991 International Workshop on Formal Verification in VLSI Design* (January 1991).
10. GOOSSENS, K. G. W. An operational semantics for a subset of the HDDL ELLA. Version 3.0 Manuscript, April 1990.
11. GOOSSENS, K. G. W. Semantics for picoELLA. Manuscript, June 1990.
12. GOPALAKRISHNAN, G. C., SMITH, D. R., AND SRIVAS, M. K. An algebraic approach to the specification and realization of VLSI designs. *Proc. 7th Int. Symp. on Computer Hardware Description Languages* (1985), 131–137.
13. GORDON, M. J. C. *The Denotational Description of Programming Languages: An Introduction.* Springer-Verlag, New York, NY, 1979.
14. GRIES, D. *The Science of Programming.* Springer-Verlag, New York, NY, 1981.
15. HOARE, C. A. R. An axiomatic approach to computer programming. *Communications of the ACM 12*, 10 (Oct. 1969), 576–580, 583.
16. ISHIURA, N., YASUURA, H., AND YAJIMA, S. NES: The behavioral model for the formal semantics of a hardware design language UDL/I. *Proc. 27th Design Automation Conference* (1990), 8–13.
17. LIPSETT, R., MARSCHNER, E., AND SHAHDAD, M. VHDL — the language. *IEEE Design & Test 3*, 2 (April 1986), 28–41.
18. LIPSETT, R., SCHAEFER, C., AND USSERY, C. *VHDL : Hardware Description and Design.* Kluwer Academic Publishers, Boston, MA, 1989.
19. MARSHALL, R. A. J., AND KAHN, H. J. A preliminary information model of VHDL. In *EURO VHDL '91 Second European Conference on VHDL Methods* (September 1991), pp. 267–275.
20. PILOTY, R., BARBACCI, M., BORRIONE, D., DIETMEYER, D., HILL, F., AND SKELLY, P. CONLAN report. In *Lecture Notes in Computer Science*, vol. 151. Springer-Verlag, 1983.
21. SALEM, A., AND BORRIONE, D. Formal semantics of VHDL timing constructs. In *EURO VHDL '91 Second European Conference on VHDL Methods* (September 1991).
22. SIEWIOREK, D. P., BELL, C. G., AND NEWELL, A. *Computer Structures : Principles and Examples.* McGraw-Hill, New York, 1982.
23. STOY, J. E. *Denotational Semantics: The Schott-Strachey Approach to Programming Language Theory.* The MIT Press, Cambridge, MA, 1977.
24. VAN TASSEL, J. P. The semantics of VHDL with VAL and HOL: Towards practical verification tools. Technical Report 196, University of Cambridge Computer Laboratory, June 1990. MSc Thesis of Wright State University.
25. WILSEY, P. A., AND DASGUPTA, S. Functional and operational specifications of computer architectures. *Proc. 9th International Symp. on Computer Hardware Description Languages* (June 1989), 209–223.
26. WILSEY, P. A., MCBRAYER, T. J., AND SIMS, D. Towards a formal model of VLSI systems compatible with VHDL. In *VLSI '91* (Edinburgh, Scotland, August 1991), A. Halaas and P. B. Denyer, Eds., IFIP TC 10/WG 10.5, pp. 6a.2.1–6a.2.12.

Part IV

SYSTEM LEVEL DESIGN AND MODELLING

Approaching System Level Design

FRANZ J. RAMMIG
Universität-GH-Paderborn, FB Mathematik/Informatik,
D - 4790 Paderborn

Abstract. After a definition of the term System Level, different system level design activities are identified: product specification and system level modeling, system level simulation, system level analysis and partitioning, embedding into a concurrent engineering environment. These activities then are discussed in more detail with special emphasis on modeling.

1. What is system level?

Today in the area of digital hardware design there is a rather widely accepted scheme of 6 abstraction levels [RA89]. This scheme is orthogonal to the different views a system is looked at. As example for this Gajski [GA87] identifies three different views: behaviour, structure, and geometry. Additional views are possible, e.g. a test view. At higher levels the behaviour view is of main interest. So in this context the levels of abstraction are discussed mostly with this view in mind. In order to make the concept of abstraction and by this the system level more visible, a bottom up presentation has been chosen. The lowest level (level 1) usually is called the **electrical level**. Here it is modelled, how electrical circuits built from resistors, capacitors, etc. behave over the time axis. This is done by a system of differential equations. I.e., both the time axis and the observable values are represented by a continuous domain. It should be noted that the geometrical view of this level is the (metric) layout which doesn't constitute an own level. The **switch level** (level 2) is the next abstract one. This level is rather well accepted in digital MOS design but makes sense in other digital designs as well. The abstraction comes from modeling transistors as ideal on/off switches and the connections in between as discrete capacitances. So the value domain is a discrete one where a value is given by a pair consisting of a logical interpretation and a strength, both within a finite domain. This abstraction introduces uncertain values. They are handled either by introducing additional "values" or by representing uncertainty by enumerating all possible values [LR83]. The time domain may still be a continuous one. Other approaches like MOSSIM [BR81] have a discrete time in mind (unit delay assumption). This leads to a concpt to model switch level circuits by finite automata. The **gate level** (level 3) has a long tradition in digital system design. It has a very nice mathematical background in Boolean algebra. However, this models only the timeless behaviour. So some additional concepts have to be considered in order to cover the time axis as well. In the ideal case the value domain is restricted to Boolean values 0,1 while

J. Mermet (ed.), VHDL for Simulation, Synthesis and Formal Proofs of Hardware, 259–276.

the time domain remains continuous. Again the problem of uncertain values forces to introduce additional values. By this in most cases the underlying algebra is no longer a Boolean one. Even the concept of different strengths is carried over from the switch level in come cases. The operators, however, are always Boolean (logical) operators. This finally constitutes this level. If the value domain is restricted to 0,1 and the time model is unit delay then the modeling concept for this level is exactly a system of Boolean equations. Further abstraction comes with the **register transfer level** (level 4). At this level a specific mode of operation is assumed. There are components that continuously observe their specific conditions. Whenever the condition of a component becomes true this component performs its specific operation. In any case such an operation may be interpreted as a transfer of data between registers where the data may be modified during the transfer. It is this point of view that gives this level its name. Abstraction here originates from implicitly underlying a specific mode of operation. In addition the elementary components used at this level are more complex (e.g. registers, ALUs, etc.) abstracting from their implementation. The value domain at this level is given by (uninterpreted) bitstrings while the timing model is the counting of clock ticks. So the time domain now has become a discrete one as well. The register transfer level is very helpful in clean synchronous designs, it forces somehow to design in this manner. This level has been studied intensively in academic institutions but is much less popular in industry. As a consequence nearly all of the numerous register transfer simulation systems (e.g. CASSANDRE [ME70, ME73, ME85], CDL [CH79], DDL [DD79], KARL [HA77]) are used rather in universities only. At the **algorithmic level** (level 5) the reactive point of view at the register transfer level is inverted to an imperative one. While at the register transfer level the system is looked at from the eyes of the individual components, at the algorithmic level the controller's point of view is taken. In contrast to ordinary algorithmic descriptions, however, concurrency plays an important role in hardware design and therefore also at this level of abstraction. Therefore highly concurrent algorithms usually are described. While at the register transfer level it is specified precisely what conditions cause operations to be carried out, it is abstracted from this information at the algorithmic level. Only the logical point of time, when an operation has to be carried out is identified. All the remaining details are hidden away by assuming the imperative mode of operation of a system. The domain of values may be freely definable but usually is restricted to bitstrings with interpretations attached. The timing model is either still a counting of clock ticks or a purely logical one. In this case simply a causality structure is assumed as in usual algorithmic languages. Algorithmic languages, have been a purely academic area for a long time. The increasing complexity of digital systems and caused by this the need for high level synthesis tools make this level more and more attractive for industry as well. VHDL [VH87] approaching this level makes it even more visible for the industrial practice. Up to now there are only very few commercial systems available for this level. DACAPO III [DA87] and partially VERILOG [TH91] and VHDL may serve as examples. Finally at the **system level** (level 6) it is abstracted from the algorithmic implementation of system's instruction sets. At this level the entire system is looked at as a set of cooperating processors. Here the term processor is used in a wider sense to denote a subsystem with an instruction set that enables it to export certain services. A usual processor is the most typical example but channels, device-controllers, etc. fall into the same class. Such a component is characterized by the functions performed by the instructions and the protocol to be used to request a service (an instruction) to be executed. In principle the initiative within such a protocol may be located at the serving device or at the requester. E.g., in the case of a usual processor th processor takes the initiative by fetching an instruction (the service it is requested to perform) from memory without explicitly being triggered to do so. In addition, to describe the components of a system and their instruction sets plus protocols, the global interaction of these semiautonomous objects has to be specified. Dependent on the kind of system to be described this may be done in a centralized manner or in a decentralized one. In the first case another highly concurrent algorithm serves to specify the global behaviour while in the second alternative in a totally distributed manner the different components decide due to certain states or events to request certain services from other components. So the system level can be interpreted as an abstraction of the algorithmic level (centralized alternative) or the register transfer level (distributed way). Both the value domain and the timing model are purely symbolic at this level. There are freely definable types with arbitrary semantics and time is interpreted only to be advanced by causality. The system level up to now is supported by very few commercial simulators. DACAPO III and VHDL are approaching this level while performance analysis tools like HIT [BE86] are addressing the system level as well.

Up to now the levels of abstraction have been looked at from the pure electronics point of view. However, very rarely pure electronic systems are built. In most cases an entire system consists of components from various domains like mechanics, hydrodynamics, aerodynamics, thermodynamics, electronics and some software running on the digital part of the electronic components. A digitally controlled car-engine may serve as a typical example where a system consisting of mechanical components, analog electronics, digital electronics running under the control of a sophisticated software have to be combined to perform an operation of high thermodynamical and mechanical complexity.

Consequently at the system level all these aspects have to be considered and the electronics part usually doesn't play the dominant role but a serving one (it's the engine's power a car driver is interested in and not the operation of the electronic controller). There may exist levels of abstraction within the modeling of non electronic parts as well but they are of minor interest for our discussion. It can be assumed that at the highest level a separation into the different domains of engineering takes place. This partitioning is a highly complex action, up to now nearly completely carried out by human decisions, based on some expert knowledge. Once the partition has been selected and the interfaces have been defined the further design can be carried out within the specific engineering domains. Now the common interfaces may get an individual interpretation and modeling in the different areas. From now on a point of view centered at the specific areas is legal. From our point of view after the step of par ioning an electronics system is obtained, providing an instruction set for software (which is of minor interest for the further steps of electronic design) and being connected to some peripheral components from a different engineering domain.

There are many degrees of freedom in these partitioning decisions. The design of a digital system may serve as example. A priory a solution providing the requested instruction set directly hardwired as hardware is as correct and as obvious as a solution providing the requested instruction set by a piece of software running on a general purpose processor which is available as a piece of hardware. And within the bandwidth spanned by these two extremes a variety of potential, correct and valuable solutions may exist. Only after one of these solutions have been selected the specification of the electronic component to be designed (if not already existing) is obtained.

2. What is system level design?

¿From the above discussion it can be concluded that partitioning seems to be the central design activity at the system level. In order to get something to become partitioned it is necessary to have a model of the entire system. Therefore modeling is essential, not only but especially at this level of abstraction. Such a model is the initial reaction to a specification of the entire system. This specification should be independent from the solution selected to find an implementation. The obtained model being the first formal document describing the system to be built some kind of verification is essential. It might be possible to proof formally (a better word is analytically) that certain aspects of the specification is met. The main activity to check the correctness of the initial model with respect to specification is simulation, i.e. experimenting whether the designer's intention is matched. System level simulation therefore is an important design activity. Besides simulation numerous additional analysis activities should be supported, like performance analysis, testability analysis, manufacturability analysis, etc. All these design tasks being performed, the already mentioned activity of partitioning can take place. It produces a couple of individual specifications for the different engineering domains. At the same time the environment for concurrent engineering has to be filled in so that during the entire design process all relevant aspects and parameters can be observed and influenced.

So the following design activities are the most important ones at the system level:

a) specification support,
b) system level modeling,
c) system level simulation,
d) system level analysis,
e) system level partitioning,
f) interaction with a concurrent engineering environment.

These activities now are discussed in a somehow more detailed way.

3. System level specification and modeling

3.1. General remarks

There is no "one and only" system level modeling method and by the same reason no "one and only" specification method. Requirements engineering being a complex area by itself we shall concentrate on the modeling aspect here. Obviously it depends heavily on how heterogeneous the possible design space is, whether an initial system model can easily be derived from a specification. In the ideal case the basic characteristics of a system to be build are fixed, only some parameters have to be supplied. Examples for such approaches are systems like DEBYS [BK91] that allow to fill in these parameters in an interactive way that is supported by a knowledge based approach. The result of this search process through a limited (but very large) design space then can be an initial model. Other examples of such an approach are generators for simple DSPs [HI85] or models of RISC processors [NA89, PS90]. This approach however seems to be promising only as long as the design space is homogeneous and limited. It is not surprising that all the examples mentioned are within a single design domain (electronic design in this case). In the general case this "ideal" way of requirements engineering will not be possible. It even will not be possible to formulate homogeneous system models. In contrary to this a multiparadigmatic approach seems to be much more promising. In such an approach for various aspects of a model different paradigms are used, for each aspect the paradigm that seems to be the most natural one or this one that is most familiar to the designer. In this section therefore simply a couple of modeling approaches shall be discussed. It should be noted that a partition of a model into parts using different paradigms doesn't imply the same partition into system components.

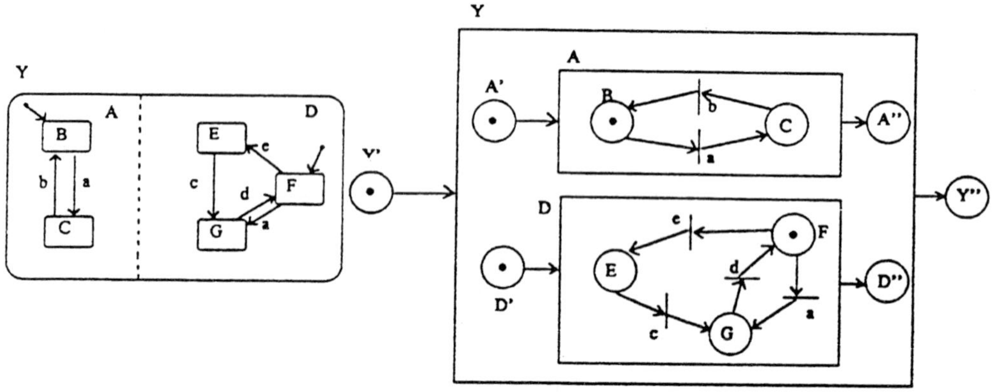

Fig. 1. Example of a statechart and equivalent S-net (from SU90)

3.2. Statecharts and structured Petri nets

Both, Statecharts and Structured Petri Nets (S-Nets), originate from finite state machines (FSM). Both add two major concepts for handling complexity: hierarchy and concurrency. And finally both are graphical approaches for system specification. Statecharts [HA 87] start from usual state diagrams of FSMs. In order to handle complexity first of all a concept of **hierarchy** is added. This is done by allowing a state to be decomposed into a FSM as well, and so on. This concept makes it necessary to introduce means for specifying in which internal state such a macrostate has to be started when activated. This may be always the same one (default initial state) or be dependent on the internal state a macrostate has been when this macrostate has been deactivated. The latter situation is supported by the so called **history mechanism** that may be extended recursively to deeper hierarchy levels. The second concept added by Statecharts to FSMs in order to handle complexity is **concurrency**. Several Statecharts now are allowed to operate concurrently. I.e. when the macrostate they are embedded in is activated more than one internal FSM is activated. Introducing concurrency always makes it necessary to introduce synchronization and communication concepts. In the case of Statecharts this is done by a **broadcasting** mechanism, i.e. by asynchronous communication.

If Statecharts may be characterized by the sequence of adding to FSMs hierarchy first and concurrency afterwards this sequence is inversed in the case of S-nets [CK81]. They start from Petri nets, i.e. the concept of concurrency has already been added to FSMs by the usual extensions introduced by Petri nets. The remaining task to find a consistent concept for hierarchy (a problem that turned out to be a complicated one) has been solved by S-nets in a very elegant way. Here each transition may be replaced by an entire S-net with a "flat" Petri net being a S-net as well. A macro-transition becomes firable by the same condition as a usual one. Internally the firing of a macro-transition means that the (always identical) initial marking is taken. Starting with this marking the local S-net becomes active and remains active as long as it is life. By this internal net becoming dead the macro-transition plays its token game in the same way a usual Petri-net transition does. Each Statechart can be simulated by a S-net and a subclass of S-nets that covers all cases relevant for practical applications can be simulated by statecharts [SU90]. So for practical applications they can be looked at as equivalent. This is interesting as Statecharts introduce hierarchy by decomposing states while S-nets decompose transitions. It depends on the special situation which concept is the more natural one. The history mechanism is missing in S-nets as originally defined by L. Cherkasova but can be added easily.

3.3. SDL

SDL (Specification and Description Language) [CC88] is a graphical language (with textual counterpart) for specification and description of systems. It is standardized by CCITT which indicates that it is especially suited for telecom applications. However it is general enough to be used in other areas as well. E.g. [GL 91] discusses the use of SDL in electronic design, relating SDL to VHDL.

SDL supports different views of a system description:

- a structural view, supported by
  = **Block Interaction Diagrams (BD)**
- a communications view, supported by
  = **Sequence Charts (SC)**, and
- a concurrent behavioural view, supported by
  = **Process Diagrams (PD)**.

The **BD**s are just usual hierarchical schematics as used at each level of electronics CAE. By identifier equality a **BD** is connected to the dynamics (the behaviour) expressed by attached **PD**s and **SC**s. **PD**s and **SC**s both describe the system's behaviour. In a **SC** the global view is stressed. It is described which communication sequences can be observed from the outside world if it is abstracted from the processes internal to the communicating objects. The inverted point of view is described by a **PD**. It abstracts from the global communication structure by just describing what messages are sent and received. On the other hand the process where these atomic communication actions play a role now is specified precisely. This multiview approach of SDL make individual descriptions very easy to understand. To get a complete understanding of a complete system, however, it is necessary to combine several subdocuments and to understand the intercorrelation. Like Statecharts SDL supports only asynchronous communication but synchronous communication (rendezvous) can be simulated easily by this concept by explicitly describing a handshake procedure. SDL is in practical use worldwide. Recently extensions towards object orientation (**OSDL**) have been proposed [MB87].

3.4. GRAPES

GRAPES-86 is a graphical language to support all aspects of system design. It has been designed by Siemens Nixdorf Informationssysteme AG [HE90] to be used for projects of high complexity. Like SDL, GRAPES makes use of a couple of correlated diagrams to describe an entire system. In GRAPES this multi-representation approach is even expanded by introducing additional types of diagrams. GRAPES is intented to combine principles of IORL (Input/Output Requirements Language) [TB84], SA (Structured Analysis) [DE78], SADT (Structured Analysis and Design Technique) [YC79] and SDL and approaches an object oriented view by looking at a system as communicating and coöperating objects. Like SDL in GRAPES the static structure of a system is represented by special diagrams. The equivalent to SDL's BD is the **Communication Diagram (CD)** in GRAPES. In addition the structure of a connection within a **CD** is further explained using an **Interface Table (IT)** that specifies the used channels and data types. The basic specification technique for behaviour is given by **Process Diagrams (PD)** with the same meaning as in SDL. There is no equivalent to SDL's **SC**s. On the other hand more emphasis is laid on the specification of data objects and data types. For this purpose **Data Tables (DT)** and **Data Structure Diagrams (DD)** are used. The counterpart to **IT**s from the process' point of view is the **Specification Diagram (SD)** where the interfaces of procedures and functions and the export interfaces of modules are specified. The entire declaration hierarchy of a system is represented by means of **Hierarchy Diagrams (HD)**.

3.5. Logic programming

Specifying a system by means of rules to be respected seems to be a natural approach. This method is not very powerful alone as long as there is no inference mechanism that allows to decide what rules are applicable and by which further knowledge can be deduced. A very general inference mechanism is given by **unification**. Therefore PROLOG [CM84] and its derivates become a useful language for specific aspects of system level specification

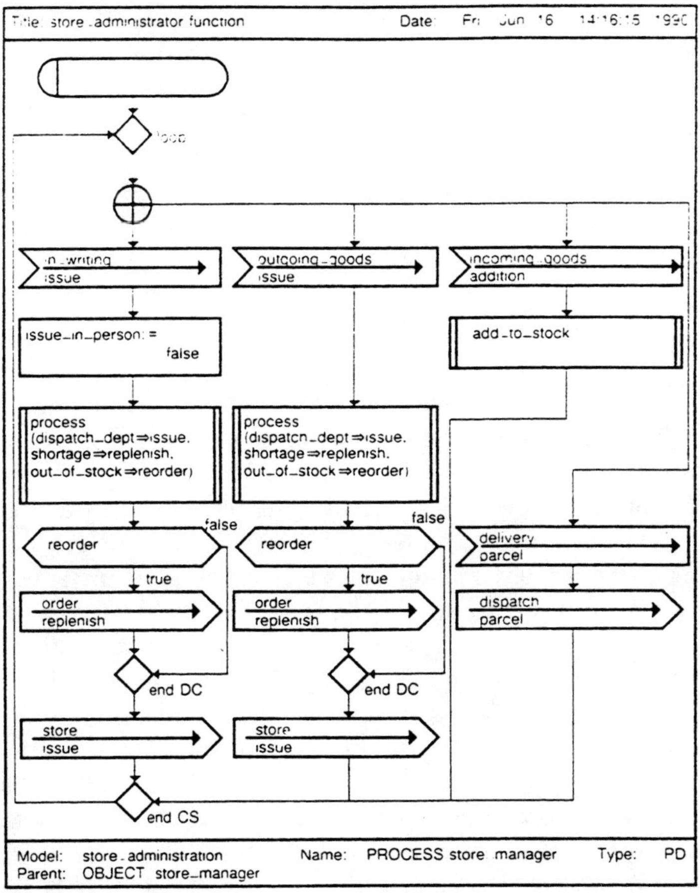

Fig. 2. Example of a GRAPES PD (from HE90)

and modeling. PROLOG is a language which is well suited to describe a calculation by means of the intended result, independently of intermediate steps. If processes where these intermediate steps are of interest rather than a final result (which may not exist) the use of PROLOG makes no sense as by the nondeterministic **backtracking** mechanism illegal intermediate states are reached that cannot be communicated to the outside. For this purpose **stream parallel committed choice** languages like PARLOG[CG84], GHC[UE85], or FCP[SH87, GK90] are well suited. FCP may serve as a typical example. A FCP clause looks like $H :- G_1, G_2, \ldots, G_n | B_1, B_2, \ldots, B_k$. H is the clause head. A goal can be unified with this head as in PROLOG. The G_i are guard predicates. The conjunction of these G_i has to be true for the clause body to be executed. If so the |-symbol (**commit**-symbol) is passed irreversably, i.e. from now on there is no back-tracking. The body predicates B_j can be interpreted as processes that are initiated concurrently by the commit operation. Fine grain communication between the concurrent processes in carried out by shared variables that might be specified as **read only variables** in certain processes. This means that such a process is not allowed to instantiate such a variable but has to wait until this has been carried out by a process that has the right to. Concurrent logic programming has been used for hardware specification [SU85, WS87]. Efficient parallel implementations are available today [GH91].

3.6. Functional programming

System modeling by functions is a very natural way with a long tradition. By systems of differential equations very complex systems can be described in a concise way. The v.Neumann-paradigm unfortunately replaces equation by assignment. As for the numerical calculation of equation systems v.Neumann computers have to be programmed this assignment point of view entered areas where it is completely inadequate. Functional modeling is very natural in any kind of continuous systems. In the area of electronics, analog devices are a typical example. But also in all kinds of engineering continuous systems are investigated and mostly described by means of differential equations. So system level modeling without supporting a functional point of view makes rarely sense. An excellent example for an approach to cover wide areas of electrical systems by functional techniques is GLASS[SE90]. Functional programming not only is a very natural approach. It also opens the way to the long tradition of mathematics. By analytical and algebraic methods formal proofs and transformations can be carried out rather simply on functional specifications while the same is very complicated in the imperative domain. Therefore it is not surprising that functional programming is playing a more and more important role in software engineering as well. This trend is increased by the observation that the parallelisation of functional models is much easier compared with imperative ones. A language that influenced functional languages in the software domain is ML [MT90] while LISP can be looked at as the classical functional language.

3.7. Object-oriented programming

If there is a paradigm that has the potential to cover most areas of system engineering, object-orientation may be the best candidate. By the principle of describing systems by a set of objects that are incarnations of object types, i.e. **classes**, the structural aspect is covered very well by the OO approach. This is achieved even better if the principle of **inheritance** is used that is present in most OO languages. By this principle hierarchies of classes can be built. Rather complicated systems of classes can be described in a concise way by this approach. Objects are not just passive entities. They offer **methods** that can be requested by other objects to be performed. Together with the principle of **polymorphism** this is a very powerful concept of abstraction. Polymorphism in this context means that in order to request a method just a message is sent to the offering object. By the nature of the message the offering object can decide how the request can be satisfied. Object orientation has a nice mathematical background by the theory of **abstract data types** (ADT) [GH78]. An ADT **D(S,E)** is given by a **signature S** and a set of **equations E**. A signature **S(sorts, ops)** is given by a set of **sorts** (domain identifiers) and a set **ops** of operations defined on these sorts. The signature specifies the syntax of the ADT. The semantics is given by the set of equations to be respected. A couple of OO languages is available today with C++ [ST86] playing a dominant role just by being inherited from C. Recently efforts are made to make concurrent OO languages more practicable [AH90]. This of course is essential if all aspects of system engineering have to be covered, as most aspects are concurrent by nature. Another area to be attacked is the specification of the protocols to be used between requesting objects and serving ones [MR89]. Traditionally in OO languages nothing is specified for this purpose. It is just assumed that any message sent from a requester is understood by the server. Technical systems behave in a completely different way, obviously.

3.8. COMPUTER HARDWARE DESCRIPTION LANGUAGES FOR SYSTEM ENGINEERING

CHDLs like VHDL have been designed to cover a specific aspect of system engineering, i.e. the design of electronic hardware. In the case of VHDL this area is further restricted to digital systems and a bandwidth of abstraction levels that reaches from partly switch level to partly algorithmic level. As dedicated languages, however, CHDLs should be the most adequate means to describe objects within their domain. As we already have identified that a multiparadigmatic approach seems to be the most promising one for system modeling. CHDLs get their natural role in this context. The approach of SPECCHARTS [VN91] is an excellent example for this idea. Here the entire system is specified by Statecharts, the fine grain functionality of the states, i.e. the operations to be performed in the states, are specified in VHDL. Another approach to add system design capabilities to VHDL is VAL [AG88]. Here additional information is added to allow to reason about provided descriptions. There are other CHDLs that offer more direct support for the system level. DACAPO III [DAC87] is an excellent example for this. In contrast to VHDL this language supports

- functional programming,
- implemented abstract data types,

- module concept,
- algorithmic concurrency

in addition to the scope covered by VHDL. Therefore it is not surprising that a couple of proposals to enhance VHDL towards system level descriptions have been made by researchers that are familiar with DACAPO III [GL90, OC90]. DACAPO III was also the basis for ODICE [MR89] an approach to introduce real object orientation, including protocol specification, into a CHDL. A perfect idea with respect of the intended polyparadigmatic approach is the concept of CONLAN [PB83] to provide a deduction system for building modeling languages instead of just defining one language. A revival of CONLAN might be a very promising way of approaching a framework for system level modeling.

4. System level simulation

By the heterogeneity of typical system level models, monolithic simulation systems hardly seem to be adequate. Therefore multisimulator-systems may be the best solution. Three major problems have to be solved when dedicated simulators have to be coupled to a multisimulator:

- data exchange between the different simulators,
- synchronization of the individual simulators,
- a unified user interface.

While data exchange can be handled rather easily by defining universal exchange formats and eventually adding information about transmitter details, the synchronization turns out to be the central problem. The simulators involved have to be kept at least synchronous enough so that at each point of time a data exchange happens, this point of time is legal for all simulators involved. There are two extrem solutions for this problem:

- the ("oversynchronizing") supervisor approach,
- the ("undersynchronizing") time warp method.

The supervisor approach being a "pessimistic" one always allows only the simulator that plans to let happen the event in the closest global future to perform this action. The method is safe but doesn't allow any concurrency. Time warping [JS85] "optimistically"assumes that the simulators can run completely independently. This assumption is OK as long as no data is transmitted to the local past of a simulator. If so, the receiving simulator has to be rolled back and resumed at an earlier local point of time. Of course when doing this, all messages sent to other simulators in the meantime have to be canceled by sending "antimessages". This may cause additional rollbacks at the receivers of these antimessages. Recently very powerful multisimulator frameworks have been reported, SICS [NI91, OC91] being an especially promising system.

Simulation at system level is mainly used to offer a workbench to the system engineer to perform experiments. Therefore special support to plan, perform, and analyze experiments have to be offered. For this purpose AI techniques have been proposed by a couple of authors [AP86, EO86, ER88, LA86, PF91, SM85]. With such techniques a knowledge based automated experimenter can be implemented that perform experiments in a goal-oriented manner. A very advanced system for this purpose has been designed and implemented by H. Pfaffhausen [PF91]. Such a system is a step to close the gap between simulation and "formal" methods.

5. System level analysis and partitioning

Besides experimenting with a modeled system various kinds of analysis may be carried out. This analysis can be based on simulation results or statically on the model itself. Performance analysis is the area where most results have been achieved. By applying queuing theory simpler situations can be solved analytically while in other cases simulation has to be applied. In this case not the system's functionality is simulated but the behaviour at the assumed queues. HIT [BE86] is an excellent example of such a performance analysis tool applicable to system level. It models systems in a very clean client/server way and offers hierarchy to cope with complex systems. Testability analysis, well understood at lower levels within the electronics domain in rarely supported at system level. Rule based approaches like this one described in [BI91] may be promising. Similar approaches may serve for an analysis of fabricatability and maintainability. In order to support a clever partitioning, analysis of similarity with predefined objects, but also within a description, is helpful. A partition where relatively many parts can be mapped into few predefined classes is a good candidate for an economic one. Therefore feature oriented retrieval operations on libraries of rather complex objects are needed. At this level of abstraction it makes nearly no difference whether the "similar" object is similar because of offering the required methods directly as hardware or by software solutions. Therefore such a retrieval system is equally helpful to search in libraries for OO programming and in libraries of hardware components. Designing such retrieval systems will be one of the most challenging tasks in system engineering of the near future.

6. Interaction with a concurrent engineering environment

When system level design is carried out all aspects of a product's lifecycle have to be considered. This coherent consideration of multiple interdependencies is called Concurrent Engineering (CE) [SS91]. Aspects to be considered include system functionality, performance, price, price/performance ratio, maintainability, fabricatability, marketing aspects, substitutability of preproducts, management cost, recycling costs, legal aspects, etc.

All these aspects have a very limited view of the entire problem and try to find local optima. To do so, influences of other aspects have to be considered, not necessarily knowing the inside of the decision-making process of the influencing aspects. Concurrent engineering has to provide the necessary information to the individual aspect-processing agents continuously and, based on a global engineering model, tries to find a global optimum as a compromise of the provided local optima. A concurrent engineering model therefore has same similarities to a system level model of an object to be designed. System level design methods therefore may influence concurrent engineering techniques. Even similar computer support seems to be probable. In any case a powerful design framework is needed for system level design with or without connected concurrent engineering. When embedded in a CE environment this framework aspect becomes even more evident. The core component of such a framework is an object oriented data base management systems (OMS) [FF91]. This OMS has to handle objects of different complexity and retrieve efficiently objects in a feature oriented way. In addition a powerful event handling and trigger mechanism has to be provided to inform and trigger tools that concurrently are working at different aspects of designing a specific system. This is the base service for CE.

Another important service of such a framework is design flow management. It manages all the versions of obtained design documents, application of tools in a concurrent or sequential manner, makes decisions about tools being best suited for a specific design, produces the reports that make the design process understandable and controllable. The third basic component of such a framework is a unified user interface management system (UIMS) helping to implement ergonomic user interfaces. Design frameworks have been identified to be the central service to be provided for concurrent engineering. They are investigated scientifically [RA87, RW91] and commercial products are emerging. One of the most advanced solution for the framework problem is the JESSI Common Framework (JCF) [KK91].

7. Conclusion

Traditionally systems have been designed by partitioning the problem into different engineering domains manually. Within these domains the system's parts then have been designed independently with the methods of different areas. The well known problem of life-cycle reduction, increasing system complexity and market pressure make a more systematic design process at the system level necessary. Such a design process has to be supported by computer based methods. Finally an integrated CE environment has to be achieved. Today the necessary base technology is evolving. So the problem can be attacked now. And it has to be attacked, especially when the european

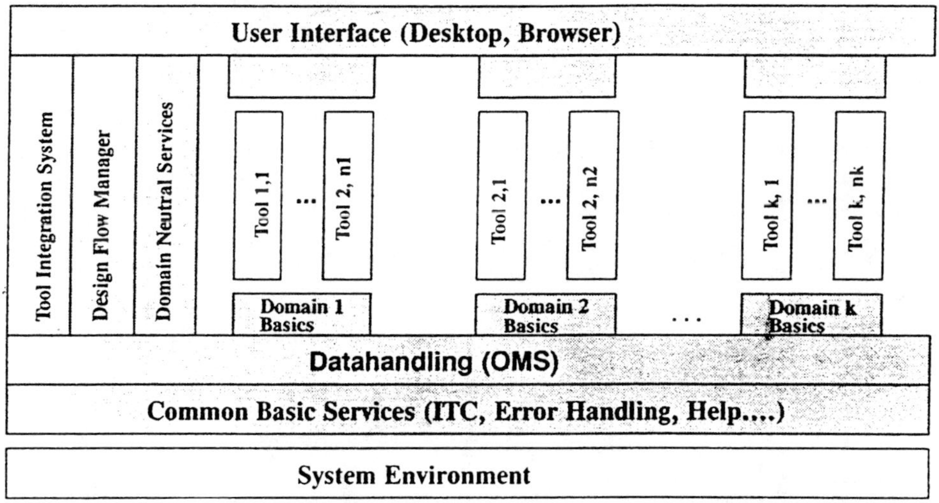

Fig. 3. Architecture of an advanced concurrent engineering environment (JESSI Common Framework)

situation is considered. Europe today is very strong in system integration and the design of highly complex systems of heterogeneous nature. To maintain this strength the understanding and application of automated system level design and concurrent engineering seems to be essential.

8. References

[AH90] J.K. Annot, P.A.M. de Haan: POOL and DOOM: the Object Oriented Approach. P.C. Treleaven (E.): Parallel Computers, Object-Oriented, Functional, Logic. Wiley, 1990

[AG88] L.M. Augustin, B.A. Gennart, Y. Huh: Verification of VHDL Design Using VAL. Proc. 25th DAC, 1988

[AP86] H.H. Adelsberger, U.W. Pooch et al: Rule Based Object Oriented Simulation Systems. In [LA86]

[BE86] H. Beilner: Workload Characterization and Performance Modelling Tools. G. Serazzi (Ed.): Workload Characterization of Computer Systems & Computer Networks. North Holland, 1986

[BI91] M. Bidjan-Irani: A Rule-Based Design-for-Testability Rule Checker. IEEE Design & Test of Computers, March 1991

[BK91] J. Bartolazzi, K. Kirsch, K. Neusinger, K.D. Müller-Glaser: Towards an Integrated Environment for Microsystem Design. In: RW91

[BR81] R.E. Bryant: MOSSIM: A Switch Level Simulator for MOS-LSI. Proc. 18th DAC, 1981

[CC88] CCITT Recommendation Z.100: Specification and Description Language SDL. AP IX-35, 1988

[CG84] K.L. Clark, S. Gregory: PARLOG: Parallel Programming in Logic. Imperial College, London, Research Report DOC 84/4, 1984

[CH79] Y. Chu: Introducing CDL. IEEE Computer, Dec. 1979

[CK81] L.A. Cherkasova, V.E. Kotov: Structured Nets. Proc. MFCS'81, Springer LNCS 118, 1981

[CM84] W.F. Clocksin, C.S. Mellish: Programming in Prolog. Springer 1984

[DE78] T. DeMarco: Structured Analysis and Systems Specification. Prentice Hall, 1978

[DA87] DACAPO III System User Manual. Dosis GmbH, Dortmund

[DD75] J.R. Duley, D.L. Dietmeyer: A Digital System Design Language (DDL). IEEE ToC, C-24, No. 2, 1975

[EO86] M.S. Elzas, T.I. Ören, B.P. Zeigler: Modelling and Simulation Methodology in the Artificial Intellegence Era. North Holland, 1986

[ER88] H.W. Egdorf, D.D. Robert: Discrete Event Simulation Methodology in the Artificial Intellegence Environment. Proc. Conference on AI and Simulation, AI Papers, 1988

[FF91] W. Fox, J. Friedrich, R. Hopp, T. Kathöfer, A. Meckenstock, D. Nolte, K. Pielsticker, G. Reitmeyer, F. Rupprecht, M. Schrewe: The Architecture of the Object Management System within Cadlab Framework. In: RW91

[GA87] D.D. Gajski: The Structure of a Silicon Compiler. Proc. of IEEE ICCD, 1987

[GH91] U. Glässer, G. Hannesen, M. Kärcher, G. Lehrenfeld: A Distributed Implementation of Flat Concurrent Prolog on Multi-Processor Environments. Proc. First International Conference of the Austrian Center for Parallel Computation, 1991

[GK90] U. Glässer, M. Kärcher, G. Lehrenfeld, N. Vieth: Flat Concurrent Prolog on Transputers. Journal of Microcomputer Applications, Academic Press, Vol. 13, No.1, 1990

[GU90] W. Glunz, G. Umbreit: VHDL for High-Level Synthesis of Digital Systems. Proc. 1st European Conference on VHDL, 1990

[GH78] J. Guttag, J.J. Horning: The Algebraic Specification of Abstract Data Types. Acta Informatica, 10, 1978

[HA77] R. Hartenstein: Fundamentals of Structured Hardware Design. North Holland, 1977

[HA87] D. Harel: Statecharts: A visual formalism for complex systems. Science of Computer Programming, 8, 1987

[HE90] G. Held (Ed.): GRAPES Language description. Siemens AG, 1990

[HI85] P.N. Hilfinger: A High-Level Language and Silicon Compiler for Digital Signal Processing. Proc. IEEE Custom Integrated Circuits Conf., 1985

[JS85] D.R. Jefferson, H.A. Sowizral: Fast concurrent simulation using the time warp mechanism. Proc. SCS Distributed Simulation Conference, 1985

[KK91] B. Kleinjohann, E. Kupitz: Tight Integration in a Hardware Synthesis System. In [RW91]

[LA86] P.A. Luker, H.H. Adelsberger: Intelligent Simulation Environments. SCS Simulation Series, 17:1, 1986

[LR83] K.D. Lewke, F.J. Rammig: Description and Simulation of MOS Devices in Register Transfer Languages. Proc. IFIP VLSI 83, North Holland, 1983

[MB87] B. Moller-Pedersen, D. Belones: Rational and Tutorial OSDL: An Object-Oriented Extension of SDL. Computer Networks and ISDN Systems 13, 1987

[ME70] J. Mermet: Définition du langage CASSANDRE. Thèse Doctorat-Ingénieur, Grenoble, 1970

[ME73] J. Mermet: Etude méthodologique de la conception assistée par ordinateur des systèmes logiques. Thèse d'Etat, Grenoble, 1973

[ME85] J. Mermet: Several steps towards a Circuits Integrated CAD System: CASCADE. North-Holland, CHDL 85, Tokyo, 1985

[MR89] W. Müller, F.J. Rammig: ODICE: Object Oriented Hardware Description in CAD Environment. Proc. IFIP CHDL 89, North Holland, 1989

[MT90] R. Milner, M. Tofte, R. Harper: The Definition of Standard ML. MIT Press, 1990

[NA89] C. Nagel: Generierung funktionaler Modelle für v.Neumann Rechnerarchitekturen. Diplomarbeit, Univ.-GH-Paderborn, FB 17, 1989

[NI91] M. Niemeyer: Simulation of Heterogeneous Models With a Simulator Coupling System. Proc. SCS 1991 European Simulation Multiconference, Juni 1991

[OC90] A. Oczko: Hardware Design with VHDL at a very high level of abstraction. Proc. 1st European Conference on VHDL, 1990

[OC91] A. Oczko, Ch. Oczko: Putting Different Simulation Models Together - The Simulation Configuration Language VHDL/S. Proc. IFIP CHDL 91, North Holland, 1991

[PB83] R. Piloty, M. Barbacci, D. Borrione, D. Dietmeyer, F. Hill, P. Skelly: CONLAN Report. Lecture Notes in Computer Science No. 151, Springer, 1983

[PF91] H. Pfaffhausen: Ein wissensbasierter Ansatz zur automatischen Durchführung von Experimenten in der Logiksimulation. Dissertation, Universität-GH-Paderborn, 1991

[PS90] B. Plorin, M. Schweins: Dialogorientierte Generierung von Mikroprozessormodellen. Diplomarbeit, Universität-GH-Paderborn, FB 17, 1990

[RA87] F.J. Rammig (Ed.): Tool Integration and Design Environments. North Holland, 1987

[RA89] F.J. Rammig: Systematischer Entwurf digitaler Systeme. B.G. Teubner, 1985

[RW91] F.J. Rammig, R. Waxmann (Eds.): Electronic Design Automation Frameworks. North Holland, 1991

[SE90] M. Seutter (Ed.): Glass: A system description language and its environment, Introduction and User manuals. University of Nijmegen, NL, 1990

[SH87] E. Shapiro: Concurrent Prolog: Collected Papers, Vol. 2, MIT Press, 1987

[SM85] R.E. Shannon, R. Mayer, H.H. Adelsberger: Expert Systems and Simulation. SIMULATION, Vol. 44, Juni 1985

[SS91] R.A. Sprague, K.J. Singh, R.T. Wood: Concurrent Engineering in Product Development. IEEE Design & Test of Computers, March 1991

[ST86] B. Stroustrup: The C++ Programming Language, Addison-Wesley, 1986

[SU85] N. Suzuki: Concurrent Prolog as an Efficient VLSI Design Language. IEEE Computer, Vol. 18, No. 2, 1985

[SU90] U. Suffrian: Vergleichende Untersuchungen von State-Charts und strukturierten Petri-Netzen. Dipl.Arb., Univ.-GH-Paderborn, FB 17, 1990

[TB84] Teledyne Brown Engineering: IORL Reference Manual. Huntsville, AL, 1984

[TH91] D. Thomas: The Verilog Hardware Description Language. Kluwer, 1991

[UE85] K. Ueda: Guarded Horn Clauses. ICOT Techn. Report TR-103, Tokyo, 1985

[VH87] IEEE Standard VHDL Language Reference Manual. IEEE IStd 1076, 1987

[VN91] F. Vahid S. Narayan and D.D. Gajski: SpecCharts: A Language for System Level Synthesis. Proc. IFIP CHDL 91, North Holland, 1991

[WS87] D. Weinbaum, E. Shapiro: Hardware Description and Simulation Using Concurrent Prolog. Proc. IFIP CHDL 87, North Holland, 1987

[YC79] E. Yourdan, L. Constantin: Structured Design: Fundamentals of a Disciplin of Computer Program and Design. Prentice Hall, 1979

Incremental Design - Application of a Software-based Method for High-level Hardware Design with VHDL

Andreas Hohl

SIEMENS Corporate Research and Development, Applied Computer Science and Software, Systems Design Automation, Munich, Germany

1. Introduction

Taking a closer look at the hardware design process it is beneficial to distinguish between "techniques", "methods", and "tools", which have to be applied to ensure an efficient design process. Because no commonly accepted interpretations of these terms exitsts in literature we give definitions here: A technique is defined to describe the "what to do aspect" aspect in contrast to the "how to do" addressed by a method. A tool, as understood here, is a computer aided assistant to support the design process, i.e. to accomplish the techniques and methods. The hardware description language VHDL is to be seen and understood as just a (programming) language, but often misunderstood as a technique or method, a misunderstanding which results in a later disillusionment of the designers, because the designers have to learn not only the language, but have also to adapt to a software-oriented design style including new techniques and methods.

The problem today is, that there is a lack of appropriate techniques and methods, as well as of tools which support (V)HDL based design, especially for complex hardware systems, the dedicated application field of VHDL. To develop techniques, methods and tools the hardware design process has to be looked at in more detail in order to find an application-oriented approach, useful to the hardware designer.

J. Mermet (ed.), VHDL for Simulation, Synthesis and Formal Proofs of Hardware, 277–290.

Therefore, this paper starts with a discussion of the hardware design process and continues with a determination of the term "hardware system" to emphasize today's and future significance of HDLs in the (hardware) design process. Afterwards the new software-based method "incremental design" is presented. In addition, a dedicated tool, "desktop design editor", is proposed which puts the theory into practise. An outlook shows the future perspectives of research in this area.

2. Hardware Design Process

There exist various kinds of "hardware design processes", depending on the application field, the design environment, e.g. tools, and the design "philosophy" of the design team. To get a common understanding of this term, a design process in VLSI design [MüDK88] is introduced, before an abstract model of the design process is presented:

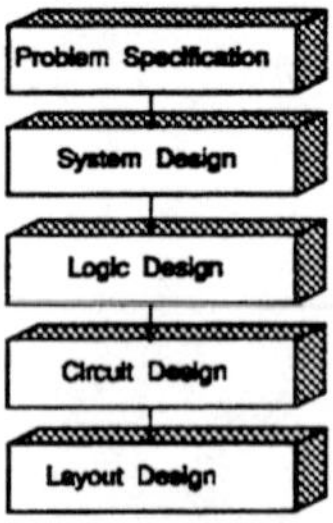

Figure 1: VLSI design process

Starting from a specification document, the designer has to build and validate models using HDLs and simulation, which describe the overall architecture of a (VLSI) system and its decomposition into modules. The logic design transforms these descriptions into a gate level description, which is translated into a circuit description before suitable geometrical layout has to be found and mask data for chip production is generated. Additional activities (validation, test data generation) accompany this process.

Today, register transfer (RT) level, as defined by the Y-Chart model [GaKu83], is often the starting point for the design process, i.e. within the system design phase the designer uses a closely hardware oriented design style and thinks in terms of solutions rather than in terms of problems, which is often true for the (problem) specification document as well. This fact has to be recognized for "new" approaches in hardware design. In the past, a move from gate level design towards higher levels of abstraction, predominantly the RT-level, is obvious. Reasons for this are the availability of synthesis tools, which automate the design process, and the use of "high-level" HDLs. The approach presented here extends this trend and addresses the phase of system design, although the design process will be an interactive one.

In case of complex hardware systems, the models to be build in a HDL often have a complexity in the range of several thousand lines of codes (loc) up to one hundred thousand loc. Experience in software development tells us the right strategy to deal successfully with such big models: top-down design. Within hardware design, common approaches lead into the same direction. The widely used Y-Chart model [GaKu83] defines a general terminology to determine the various levels of abstraction and views, which can be identified within an arbitrary design process. But the focus is rather on the design process itself.

In [Ramm89] a model of the design process is discussed in detail. At each level of abstraction [Ramm89] defines the design activities of "generation" and "checking" of a HDL model within a hardware system design. Input at various levels are expressed in "design languages" (in a more general sense) including HDLs, natural language text, graphics, timing diagrams, etc.. At each level, design documents and the input to the following lower level of abstraction are the output.

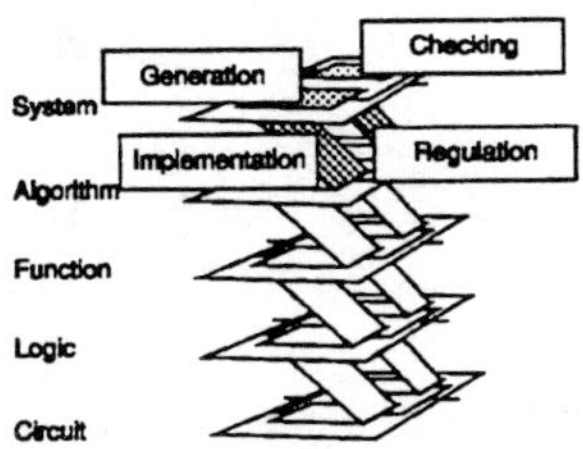

Figure 2: Microscopic view of the design process

The further design activities of "implementation", e.g. from RT-model to a gate model, and "regulation" [cf. Ramm89], e.g. backannotation of timing parameters, complete this model for a multi-level design, which results in a microscopic view of the design process. In the end, a macroscopic, at each level of abstraction, and a microscopic model of the design process is defined, but the aspect of method as understood here is not addressed. Especially the aspect of introducing software-based approaches, e.g. CASE tools and methods, is not addressed.

3. Hardware Systems

Having a closer look at "hardware systems" it becomes obvious that these systems contain non-hardware parts, e.g. software parts, as well. Using HDLs such as VHDL it is clear as well, that (V)HDL models are able to describe software parts. In order to agree on a strategy for system design and to anticipate future applications and needs of developed approaches, the term "hardware system" has to be defined in more detail.

system "Anything we choose to regard (a) as an entity and (b) as comprising a set of related components. In computing the word is widely used with many shades of meaning. Most commonly, however, it may refer to a related set of hardware units, or programs, or both. ..." [Dc90].

Thus, a (hardware) system can be characterized by an architecture, the decomposition into modules, an interface and a behavior. The outside world can perceive its behavior by observing the interface of the system.

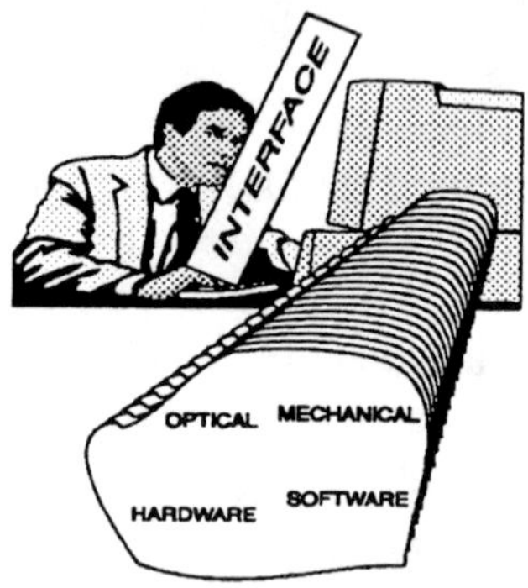

Figure 3: Schematic view of a system

Up to this point, the view of a system is not limited to hardware systems, but is applicable for software systems and others as well. In order to determine what is meant by "hardware", a bottom-up view can be applied to a system. Hardware components, of which a hardware system is built, include digital components as integrated circuits (IC) or application specific ICs (ASIC), digital components as printed circuit boards (PCB), which contain ICs/ASICs, wires etc., analog components, optical components, and mechanical components. In general, the complete (hardware) system needs firmware and software components as well to show the required behavior.

Commercial hardware design environments of today mainly support the design of digital hardware systems or components. A trend is to be recognized to support the design of PCBs in a more adequate way and to pay attention to a mixed-mode, digital and analog, or pure analog design [ReWe89a, ReWe89b].

In a first step, the method of incremental design is applied to digital hardware systems. The strategy is to introduce this advanced modeling method to the hardware designer, who wants to address system design, and to develop further methods on top of it. Moreover, incremental design can be utilized next to design more complex system, e.g. including hardware and software components.

4. Incremental Design

The new approach presented in this section is influenced by software development approaches [Bisch89, Vonk90], essentially by rapid prototyping, on the one hand and various experiences in hardware design using HDLs [e.g. Hohl86] on the other hand. According to the hierarchical definition of the terms technique, method and tool the starting point is a top-down strategy, which requires various methods tackling specific design problems within the design process and has to be supported by tools to be developed.

Technique: Top-down Design

The author defines the term "top-down design" in a new way and chooses this strategy to integrate the method of incremental design into the design process. The term top-down design means in this context that the design process is based on the software life cycle [cf. Vonk90] on the one hand and on the microscopic design process model [Ramm89] on the other hand. The chosen HDL here is VHDL, but the underlying concept is not limited to VHDL.

Method: Incremental Design

As emphasized in the preceding sections, (hardware) design processes as well as (hardware) systems are of various kinds. To deal with this challenge, the method of incremental design defines basic design activities, valid for (almost) all kinds of applications, application-specific design constrains, and design options. These basic activities, which define the incremental design method, are discussed next.

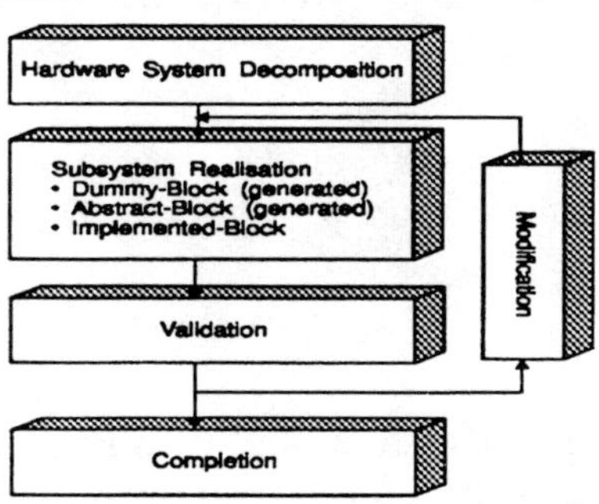

Figure 4: Incremental Design

Design Activities:

The basic design activities, outlined in Figure 4, are the following:

1) Hardware System Decomposition

Similar to structured analysis in software engineering, the hardware designer has to decompose the hardware system into subsystems and to define their interfaces.

2) Subsystem Realization

For each subsystem a behavior, i.e. the function the subsystem is to perform, has to be specified. Therefore, the main idea of the incremental design method is to provide the designer with a complete HDL model, i.e. ready for simulation, where submodels are realized only to a certain degree. Three kinds of realization degrees can be distinguished:

a) Dummy-Block:

Only an interface description of the submodel is present. The body, where the functional and timing behavior is described, remains empty. The option "generated" stands for generated by a tool and is described later.

b) Abstract-Block:

Here, the dummy-block is enhanced by an abstract description of the functional and timing behavior within the body. The term "abstract" is not necessary related to the term "levels of abstraction", as described in the Y-Chart model. An abstract-block can be described on different levels of abstraction, even with different views (behavior or structure).

c) Implemented-Block:

At this stage, the body of the submodel is described in detail, which is in general the designer's tasks. As mentioned before, the designer can use the full range of description levels as defined by the Y-Chart model.

3) Validation

The validation of the resulting model, or preliminary versions thereof, which contains submodels of mixed degrees of realization will be done by simulation.

4) Modification

The loop of realisation, validation and modification is continued as long as the validation fails.

The method of incremental design reduces the design time drastically. By the step of completion, the designer gets a complete and validated model of the hardware system.

Design Constrains:

The second issue, determination of design constrains, can be handled if the possible design constrains are classified into three categories:

1) Abstraction Levels Constrains

The designer has to determine the top-level and the further addressed levels of abstraction of the design process. With respect to the widely used Y-Chart model and for practical reasons, the following levels could occur as a top-level:

a) Architecture or System Level:

At the system level, the system is decomposed into components and the environment is described. Characteristics, e.g. performance, are specified and the timing behavior is given in a global manner (e.g. asynchronous communication between processes).

b) Algorithmic Level:

The basis for the description at the algorithmic level is the partitioning of the system from the previous design step. Each component of the system is implemented by an algorithm. By definition of the term "algorithm", the algorithmic level is closely related to the behavioral view, although view-dependent design objects can be characterized. The timing behavior is described in a global manner as at the system level.

c) Functional Block Level or RT-Level:

The decomposition of the components of the system continues. The imperative description style, similar as known from general programming languages like PASCAL or C, of the algorithmic level is replaced by a reactive description style, i.e. the point of view in the description is the view of the design objects. Timing behavior is explained in a more detailed manner, e.g. cycles of process execution. This level is often refered to as register-transfer level, although this term implies a mix of levels of abstraction and views.

2) Constrains from the Design Process

After finishing the HDL model generation, either interactively or automatically, the model is to be validated, often using simulation, and to be implemented at a lower level of abstraction, using synthesis tools. Today, these tools require a specific description style, especially synthesis tools, to be efficient or even correct.

3) Hardware System Constrains

The design object itself implies of course several design constrains, e.g. the use of specific component libraries to design a PCB, specific timing behavior in case of synchronous and asynchronous designs and the complexity of the hardware system and the HDL model to be build (small, large, many).

Design Options:

The third issue to be addressed is the determination of design options, which can result from the expertise of the designer, the quality of the tools and the complexity of the design process. Two basic options can be identified:

1) " Application-oriented" Design:

In an ideal design process, the design object should be in the center of interest and not the design environment. Unfortunately, this is not the case today in reality, but a step towards the ideal is to support an application-oriented view in the design environment. For instance, if the designer wants to build a HDL prototype of a CPU, the design system advices and supports him or her to build a (VHDL) entity first and to use processes to describe a finite state machine. This is meant by application-oriented design.

2) "Language-oriented" Design:

The second option, language-oriented design, refers to the fact that design environments are not able to handle all kinds of designs in an efficient way. Therefore, to give the designer direct access to the process and to bring in his or her expertise, the used HDL must optionally be transparent to the designer.

Incremental design can be supported by a "desktop design editor". The concept of such a tool is presented next, enhanced by some examples.

Tool: Desktop Design Editor

The method of incremental design describes the basic design activities to be handled and in addition optional design constrains. The focus is now on the hardware system decomposition and the subsystem realization.

1) Hardware System Decomposition:

In general, the hardware designer has to identify the subsystems. An expert system like component can assist in doing this task, based on empirical knowledge. In the prototype design system being developed, the desktop design editor (DDE) displays a form to be filled in from which VHDL code is generated and which is used for graphical display as well. To allow an unique treatment of hierarchical subsystems, they are refered to as "design objects" in the following.

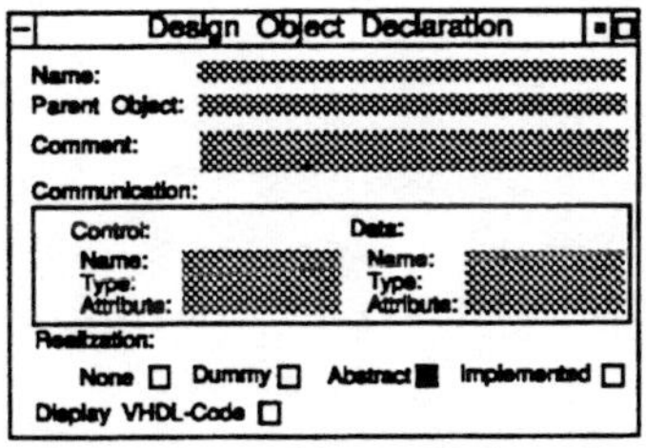

Figure 5: Design object declaration form

At first, a design object (DO) is assumed to be equivalent to a VHDL entity. The mandatory name entry determines the entity name; parent object identifies the position within the hierarchy; comments helps to explain the function of the design object; the communication part identifies the interfaces of the entity and is separated into a control and data part; interfaces can be attributed, e.g. by unique identifiers that allow automatic mapping of interface values in mixed-level design. The realization part induces the generation of an architecture, which is explained in the following design activity, subsystem realization. To allow transparent VHDL-codegeneration, a display option can be switched on.

The assumption that a DO is equivalent to a VHDL entity is not mandatory at all, because the communication part of the form is variable and can describe DOs to be mapped to (VHDL) subprograms, processes, blocks or configurations, even an arbitrary translation to a programming language program is conceivable.

2) Subsystem Realization:

Each subsystem is defined as a DO, which is characterized by a name, a schematic symbol and a realization, e.g. as a VHDL model. DOs are parameterizable and stored into a data base. The designer builds the hardware system by putting down and connecting DOs with a schematic or textual editor. DOs could be library components, e.g. ALU, MUX, register or adder, or more abstract components, e.g. FSM or communication block. The latter one is very general in its application and therefore introduced to describe the idea of subsystem realization to a certain degree.

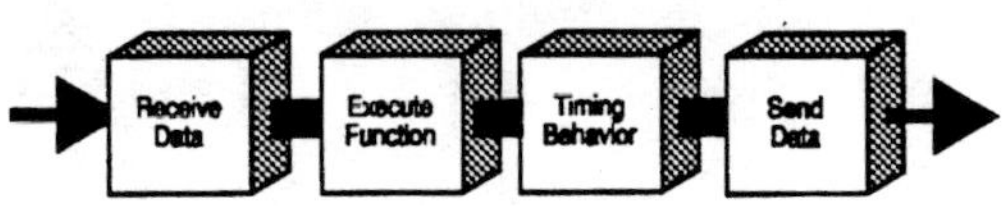

Figure 6: Communication Block

A communication block (CB) is characterized as a black box, with an input and output interface, which implements the following basic functions:

a) Receive Data

Data is taken from the inputs, optionally modified (e.g. format conversion), and stored into internal variables.

b) Execute Function

This function of a CB "simulates" the execution of the DO. Depending on the degree of realization, the "real" (e.g. internal structure) hardware function is described or the resulting behavior to be observed by the CB's environment.

c) Timing Behavior

The timing behavior is separated from the function. In a simple case it may be described with a VHDL wait statement, wait for a specified period. More complex timing conditions, e.g. asynchronous timing behavior, are conceivable as well.

d) Send Data

The resulting data from the function is send to the environment. In some cases it may be necessary to add timing information to this block.

A CB is implemented by the following VHDL entity:

```
entity communication_block is
        generic (Generic List);
        port (Port List);
end communication_block;

architecture template of communication_block is
Architecture Declarative Part
begin
        receive_data:        Concurrent  Statement;
        execute_function:    Concurrent  Statement;
        timing_behaviour:    Concurrent  Statement;
        send_data:           Concurrent  Statement;
end template;
```

Figure 7: VHDL Template of a CB

Example I2C-Bus-Interface:

To demonstrate the design activities, a hardware system containing a processor, mapped via a processor interface and an I2C-Bus-Interface to an I2C-Bus is used.

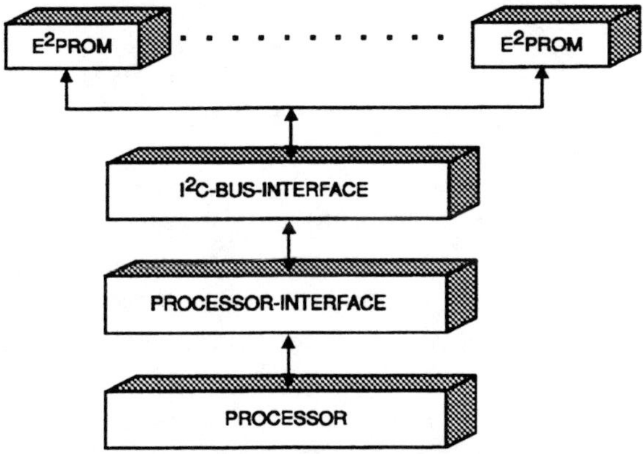

Figure 8: Hardware System

Within the system decomposition phase, all subsystems [cf. Figure 8] and their interfaces are to be determined. The specification documents define the interfaces at the RT-level, i.e. as signals and vectors of signals. To focus on the subsystem realization as an abstract-block, it is assumed that the interfaces are implemented as specified, the realization of a dummy-block only results in an empty architecture body and the implemented-block will result in a RT-level description, which can be synthesized using commercial synthesis tools. The I2C-Bus-Interface subsystem is realized at this level using a CB as described in the following:

```
entity i2c_bus_interface is
        port(   rsh:        in bit_vector(40 downto 39);
                dip:        in bit_vector(7 downto 0);
                ....
                dop:        out  bit_vector(7 downto 0));
end i2c_bus_interface ;

architecture i2c_abstract_block of i2c_bus_interface is
        ...
        procedure preceive_data(...) is ...
        procedure pexecute_function(...) is
        ...
        case i2c_opcode is
          when select_e2prom => ...
          when  write_e2prom => ...
          when read_e2prom => ...
          when others => ...
        end case;
        ...
        end pexecute_function;
        procedure ptiming_behavior(...) is ...
        procedure psend_data(...) is ...
```

```
begin
        receive_data:         preceive_data(...);
        execute_function:     pexecute_function(...);
        timing_behaviour:     ptiming_behavior(...)
        send_data:            psend_data(...);
end i2c_abstract_block;
```

Figure 9: VHDL Model (abstract-block)

The behavior of the I2C-Bus-Interface is described within the procedure pexecute_function by implementing the different bus operations defined by the I2C-Bus-Protocol. Because of the asynchronous communication between this subsystem and the processor, all activities are sensitive to the occurrence of data events. The other subsystems are described in a similar way. Within the next step, the RT-level description of the I2C-Bus-Interface subsystem is performed. The complete model at the "abstract-block level" can serve as a test bench during the succeeding design phases.

The DDE being developed supports the design activities presented and allows thereby for an efficient design process.

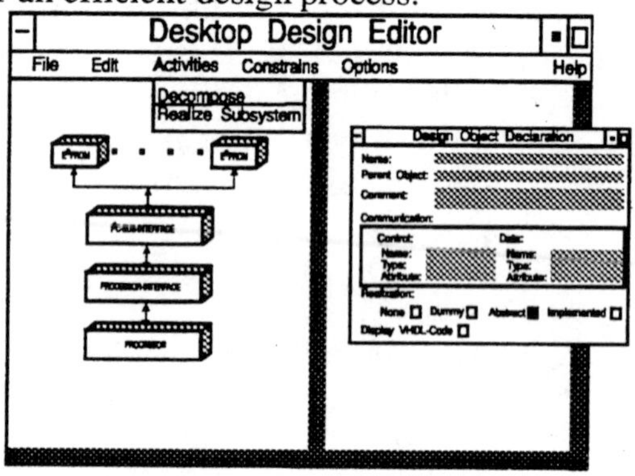

Figure 10: Sketch of the DDE

The DDE is OSF/Motif based; it allows in a first version a textual manipulation of VHDL models, supports the design activities of system decomposition and subsystem realization presented before and permits the specification of design constrains and design options.

Integration into the Design Process:

Hardware design is a complex process with many relationships between the different phases. For a new technique or method to become successful, they have to be well-integrated into this process. Two aspects have to be recognized: Tool-integration and relationship to other applied techniques or methods. The aspect of tool-integration is considered by design for integration of the "design system", where the DDE is a part of. As other tools, e.g. synthesis, are applied in succeeding design steps, coexistence of incremental design and the DDE with other techniques or methods is an issue that can not be neglected. Therefore, the templates or the generated (VHDL) code can optionally be restricted to a given subset, accepted by the synthesis tool. In addition, an optional synthesis tool dependent description style supports this issue, i.e. the description style is conformant with the tool dependent VHDL "cookbook".

5. Conclusion

The software-based method incremental design is presented which supports the system design step, where high-level HDLs are applicable. A prototype of the DDE is being developed and further design experience will be gained in applying the method and the tool in practise. Further methods, mainly prototyping approaches, will be build on top of it.

Acknowledgements:
The author likes to thank Prof. B. Reusch, University of Dortmund, and Dr. U. Lauther, SIEMENS, for fruitful discussion.

References

[Bisch89] Bischofberger, W.R., Prototyping-Oriented Incremental Software Development, Paradigms, Methods, Tools and Implications, Ph. D. Thesis, University of Linz, 1989.

[Comp74] Various papers in IEEE Computer, Volume 7, No. 12, December 1974.

[Comp77] Various papers in IEEE Computer, Volume 10, No. 6, June 1977.

[Dc90] Dictionary of Computing, 3rd Ed., Oxford, 1990.

[Dt90] D&T Roundtable Behavioral Description Languages Part 1: Are Designers Benefitting?, IEEE Design & Test of Computers, February 1990.

[GaKu83] Gajski, D., Kuhn, R., Guest Editors' Introduction: New VLSI Tools, Computer 16(12): Dec. 11-14, 1983.

[GaSa] Gane, C., Sarson, T., Structured Systems Analysis: Tools & Techniques, IST Databooks, New York, 1977.

[GiSh85] Giloi, W.K., Shriver, B.D., An Overview of the Computer System Design Process, in: Methodologies for Computer System Design, Editors: Giloi, W.K., Shriver, B.D., North-Holland, 1985

[Hohl86] Hohl, A., PIPECAD, an application-oriented CAD-Tool for the Design of pipelined Processors (in German), Masters' Thesis, Dept. Computer Science of Dortmund University, 1986.

[Hohl90] Hohl, A., HDL for System Design, in: Advanced Information Processing, Proceedings of a Joint Symposium Information Processing and Software, Systems Design Automation, Schwärtzel, H., Mizin, I. (Eds.), Moscow, June 5/6, 1990, Springer 1990.

[HöNe86] Hörbst, E., Nett, M., Schwärtzel, H., VENUS Design of VLSI-Circuits (in German), Springer 1986.

[McPC88] McFarland, M.C.; Parker, A.C.; Camposano, R.: Tutorial on High-Level Synthesis, Proceedings of the 25th Design Automation Conference (1988).

[MüDK88] Müller, J.A., Dittrich, K.R., Kotz, A.M., Design Management Support by Advanced Database Facilities, in: Rammig, F.J. (Ed.), Proceedings of the IFIP WG 10.2 Workshop on Tool Integration and Design Environments, North-Holland, 1988.

[NaSp90] Navabi, Z., Spillane, J., Hardware Generation from a Synthesis Subset of VHDL, Proc. of the VHDL Users' Group, Boston, MA, April 1990.

[NeSa87] Newton, A.R., Sangiovanni-Vincentelli, L., CAD Tools for ASIC Design, Proceedings of the IEEE, Vol. 75, No. 6, June 1987.

[Newt90] Newton, A.R., VLSI-Based Systems Design Challenges in the Early 1990s, in: Advanced Information Processing, Proceedings of a Joint Symposium Information Processing and Software, Systems Design Automation, Schwärtzel, H., Mizin, I. (Eds.), Moscow, June 5/6, 1990, Springer 1990.

[NuMe89] Nurie, G.M., Menchini, P.J., VHDL Model Portability, High Performance Systems, July 1989.

[PBBD83] Piloty, R, Barbacci, M., Borrione, D., Dietmeyer, D., Hill, F., Skelly, P., CONLAN Report, Lecture Notes in Computer Science, No. 151, Springer 1983.

[Ramm89] Rammig, F.J., Systematic Design of Digital Systems (in German), Teubner, Stuttgart, 1989.

[ReHe88] Reusch,B., Herms, J.A., Design and Simulation of Microelectronic Systems (in German), CAD-CAM Report No. 11, 1988.

[ReHe90] Reusch,B., Herms, J.A., Digital Simulation, Part II, Analysis and Testpattern Generation (in German), in: Forschungsberichte des Fachbereiches Informatik der Universität Dortmund, Bericht Nr. 340/1990.

[ReWe89a] Reusch,B., Weinerth, H., Key Technology Microelectronic, Part 9: Future Trend of Computer Aided Design Methodology for IC Design (CAD) (in German), Elektronik, 1989.

[ReWe89b] Reusch,B., Weinerth, H., Key Technology Microelectronic, Part 19: Future Trend of Computer Aided Design Methodology for IC Design (CAD) (in German), Elektronik, 1989.

[Sang85] Sangiovanni-Vincentelli, A.L., An Overview of Synthesis Systems, IEEE 1985 CICC.

[VHDL87] IEEE Standard VHDL Language Reference Manual, IEEE Std 1076-1987.

[Vonk90] Vonk, R., Prototyping. The effective use of CASE Technology, 1990.

[WaTh85] Walker, R.A., Thomas, D.E., A Model of Design Representation and Synthesis, Proceedings of 22nd Design Automation Conference, IEEE, 1985

Introducing CASCADE control graphs in VHDL

Claude Le Faou, Jean Mermet

ARTEMIS-IMAG, BP 53X, 38041 Grenoble cedex, France

1 . Introduction

One basic principle of hardware design is a clear separation of control and operations.

Not all Hardware Description Languages reflect in their concepts this separation. With APL used for describing the architecture of system 360 [FIS64], everything was mixed. With VHDL some *if* or *case* statements clearly address control as well as *guarded* assignements. But a clear separation requires either some overconstraint on the language or some specific language constructs performing a one to one mapping with control concepts (ex. Finite State Machines).

In CASCADE [Me85], an explicit control graph is used in behavioural description at system level or at RT level. At lower levels (gate or switch) these concepts are irrelevant because they have already a logic network implementation.

The CASCADE control graph can either be described textually or graphically. An automatic translation is performed between graphics and text (and vice versa). We shall not here describe the textual representation because CASCADE is not candidate as a standard language. But we think that it would be user friendly and often much more efficient to describe control properties of a system by means of a graph like the CASCADE control graph. We shall give an automatic VHDL translation which will guarantee through a simple preprocessor full compliance to the standard.

The same control structure can apply at two different levels whether or not concurrency exists in the control. Possible presence of asynchronous control and parallelism characterize system level. But the same graph, used under constraints can describe the FSM. used at RT level whose transitions from state to state are synchronized. Any intermediary combination can apply (like for example a hierarchy of concurrent FSM), as well as either statecharts, or S.Nets [Ra91].The CASCADE control graph was inspired by generalized Petri Nets like for example LOGOS [Ro72].

J. Mermet (ed.), VHDL for Simulation, Synthesis and Formal Proofs of Hardware, 291–307.

2 . The Control graph

The graph is a directed dual graph with two categories of nodes : "places" and "transitions". A place can be connected only to a transition and vice-versa. Nevertheless, in order to make the graph more compact, the graphical representation will allow certain regrouping of elementary transitions in order to create "composed" more powerful transitions. We shall define the semantics of these composed transitions in two ways :

- The operational semantics reflecting their behaviour during simulation. The VHDL translation will be one way to express the operational semantics.
- The static semantics, by means of recursively defined graphs, starting from a few very simple primitives.

2.1 PRIMITIVES

a. The place

The place (represented ○), is a primitive which has two states : *token*, *empty*.

A token can enter a place through an edge directed towards this place, or leave it through an edge directed to a transition.

We are using two axioms, which generate error messages when violated:

- A token cannot be sent to a place which is not empty.
- A token cannot go to two different transitions.

A static verification that the first axiom is satisfied in a given graph is not always possible. Acces conflicts to a place are precisely what a designer wants to discover through simulation. A static verification that the second axiom is verified is easy just looking at the topology of the graph.

In general we would consider as reasonable to detect statically places with more than one output edge. In particular cases however we shall use this possibility on purpose to describe, for example, mutually exclusive (in time) transitions.

b . Elementary transitions

A transition fires when given conditions are satisfied. When an elementary transition fires, all tokens contained in input places are swallowed by the transition which loads with tokens certains of its output places.

b1 AND

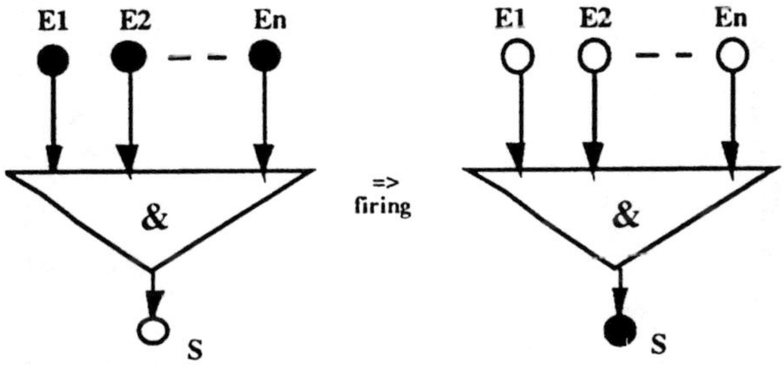

When the transition fires, the state of the primitive moves from left to right in the above figure.

S <= token when (E1=token) and (E2=token) and ... and (En=token);

E(1 to n) <= empty;

b2 SINK

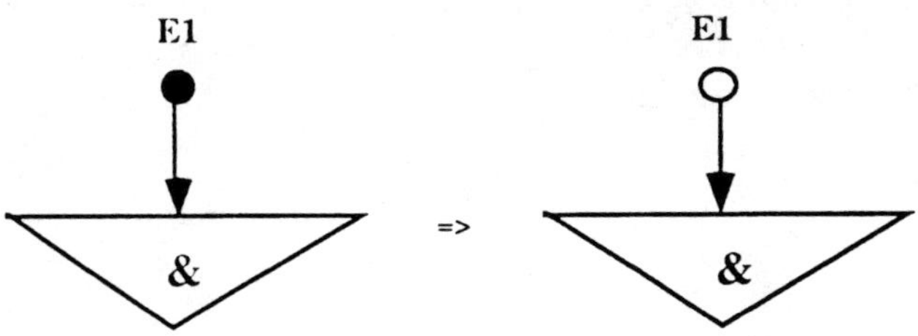

A one input AND with no output is a sink for tokens

b3 Token multiplicator

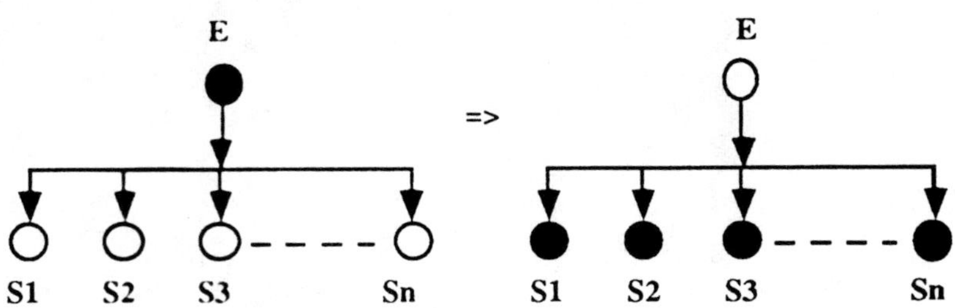

if E=token then S1 <= token ; S2 <= token ; S3 <= token ; ...;

Sn <= token ; E <= empty; endif ;

b4 TEST

Let B be a boolean

Before firing After firing, if COND if not COND

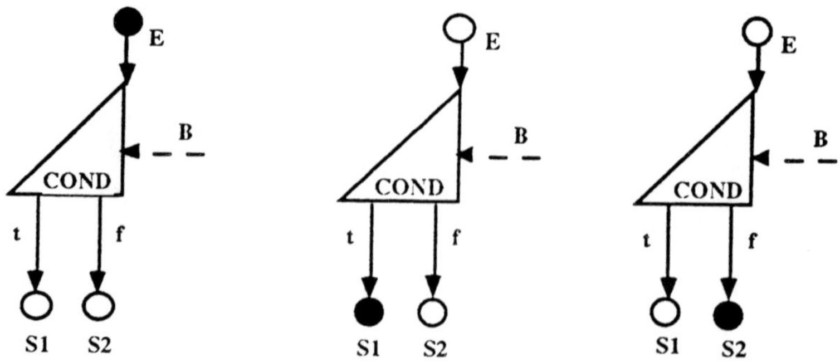

If E=token then S1 <= token when B='1'; S2 <= token when B='0';
E <= empty ; endif;

- let B be a place

We shall represent in the same way the test that a place B contains or not a token. It is easy to imagine that there exists in the operative part of the described system, a simple process associating a boolean to each place which is true if a token is in the place, else false.

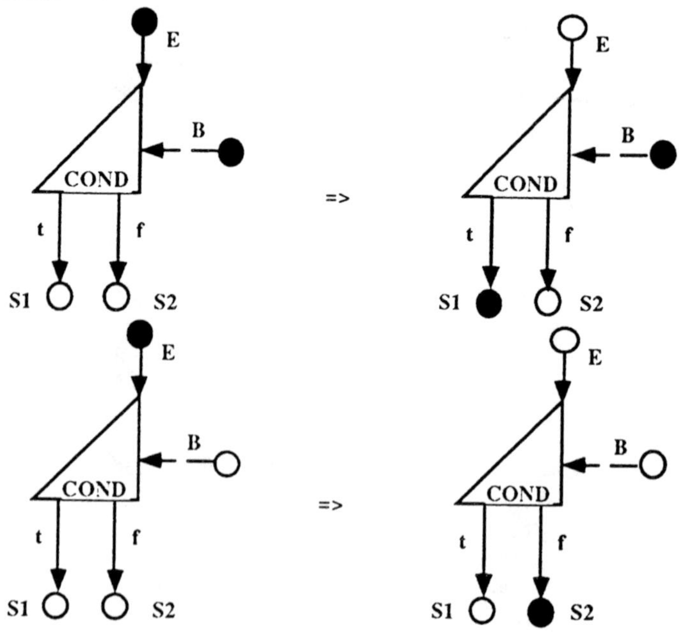

If E=token then if B=token then S1 <= token ; else S2 <= token ;
endif; E <= empty ; endif ;

b5 DELAY

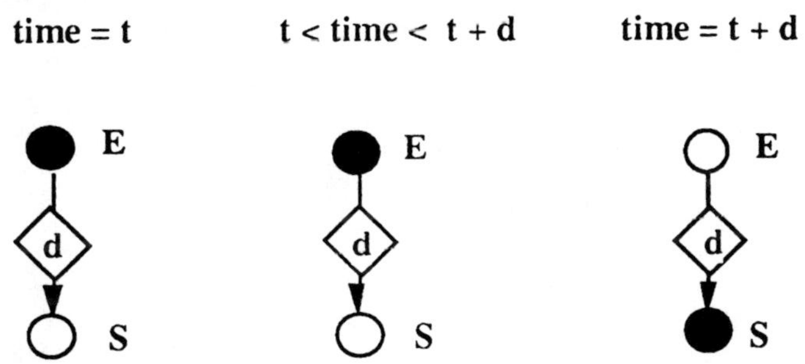

S <= token after d when E=token ; E <= empty after d when E=token ;

c Composed primitives

c1 Multiple output AND

If all inputs contain tokens then all outputs are loaded

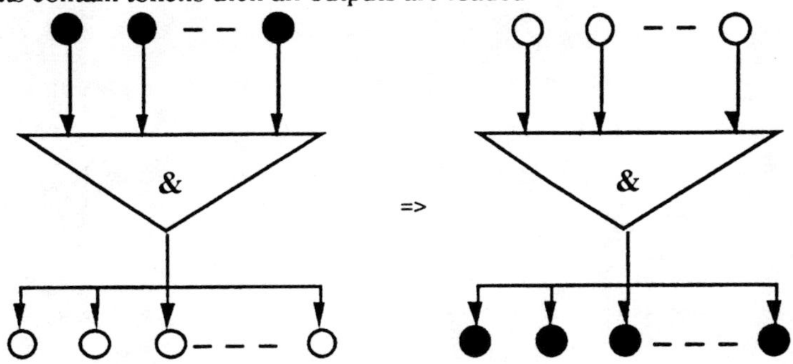

c2 SYNCHRO

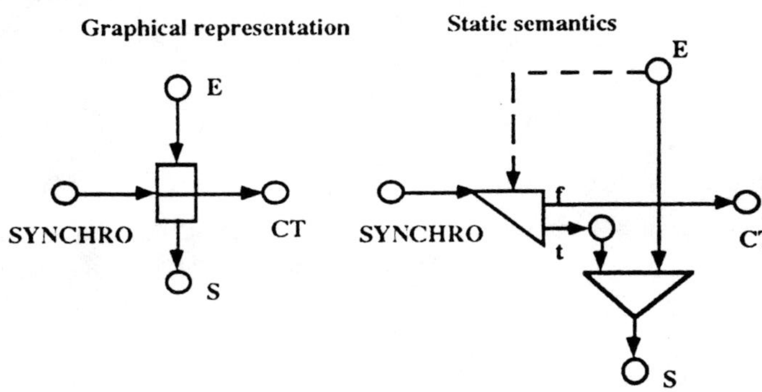

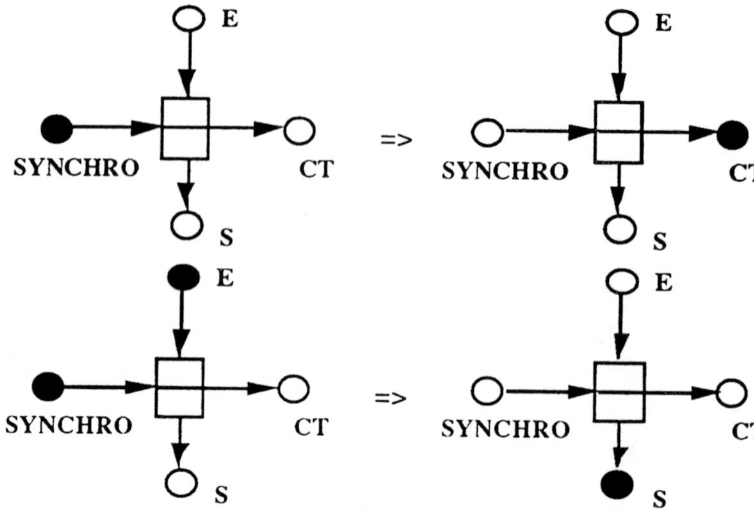

When a token arrives in E it is memorized until a token arrives in SYNCHRO.

When a token arrives in SYNCHRO
if (E=token) then (S <= token, SYNCHRO <= empty, E <= empty)
else (CT <= token, SYNCHRO <= empty)
In all cases SYNCHRO cannot memorize a token contrarily to E.

c3 STATE

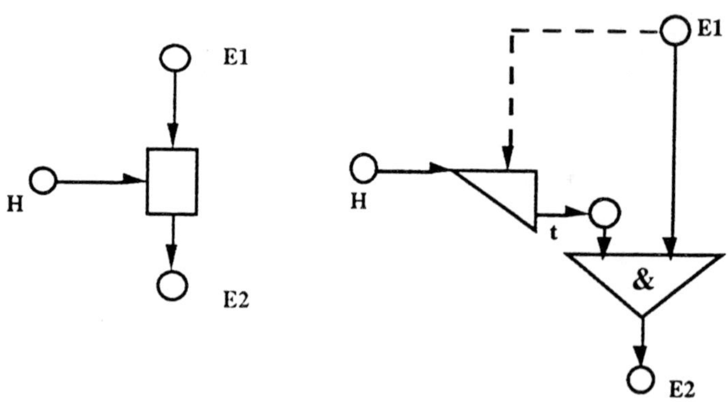

In this case one output (CT) is withdrawn

If E1 is a state in a FSM, then a token in E1 means that the automaton is in a state E1. The occurence of a clock H, on input SYNCHRO causes the transition of the automaton from E1 to E2

c4 SELECT

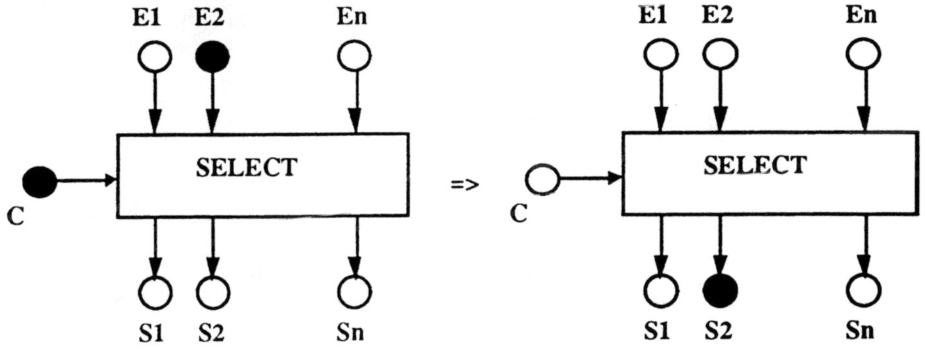

When a token arrives in C :

If there is no token in the inputs, the token C is memorized in C

If there is a token in only one input Ek, output Sk is loaded, Ek and C are cleared.

Static semantics :

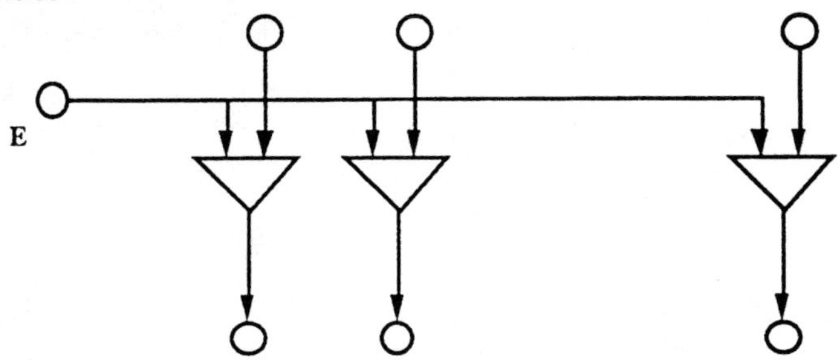

If there are tokens in more than one input then it is an error.

Here is an example of using the topology of the graph (place connected to several transitions) to verify the constraint that a single input is allowed to store a token.

c5 PRIORITY

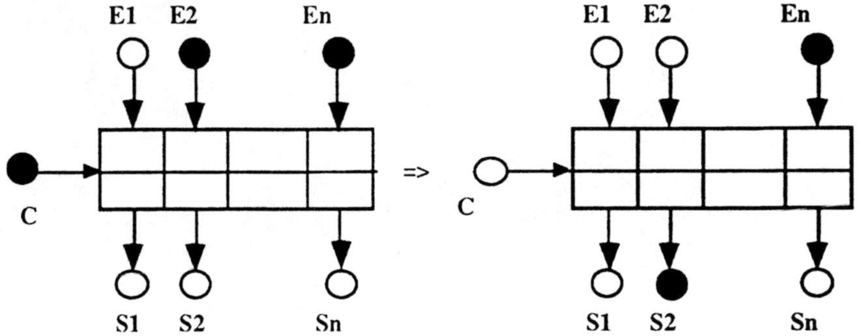

When a token arrives on C, if there are tokens in several inputs the first token met, when scanning input from left to right, is loaded in the corresponding output.

The tokens which could be in other inputs are memorized.

PRIORITY is composed of cascaded SYNCHRO primitives.

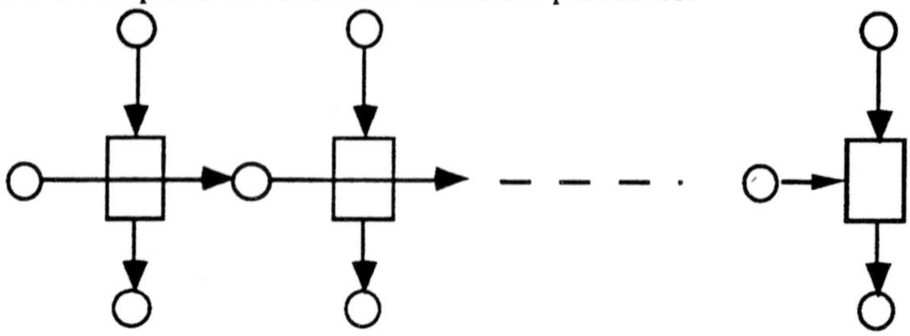

Static semantics : two cases must be addressed :

The token arriving in C is lost if all inputs are empty (normal case)

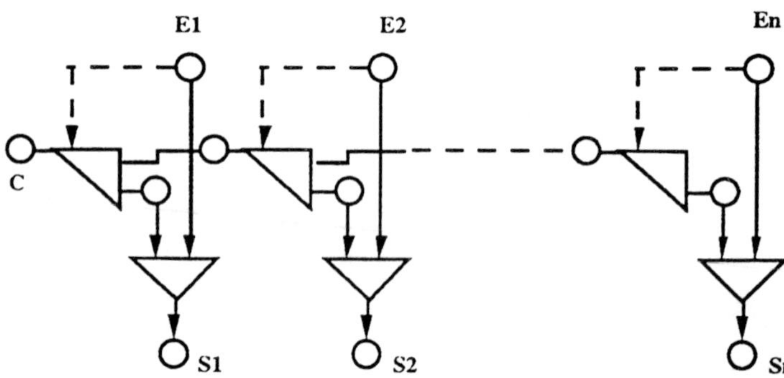

The token arriving in C is fed back in C if all inputs are empty.

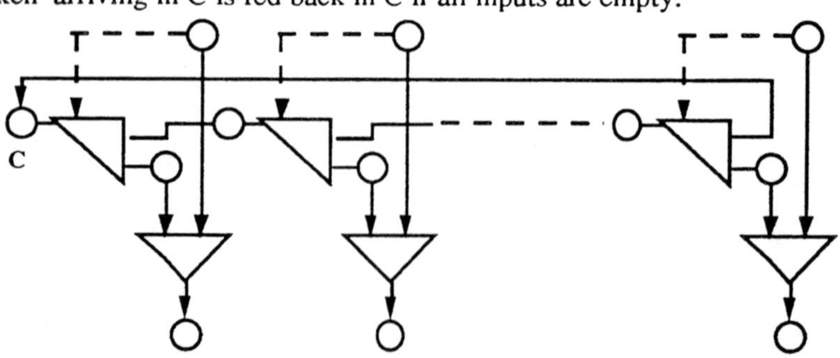

We shall graphically represent it :

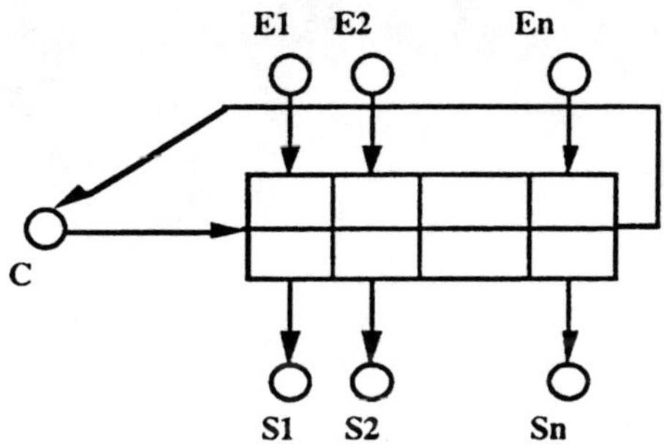

Examples of mutually exclusive transitions

As we have already seen in C4, we can use the topology of the control graph to model mutually exclusive operations. Lets call C1 and C2 the control tokens of two transitions on which we want to impose that they must never be enabled at the same time.

<u>Example</u>

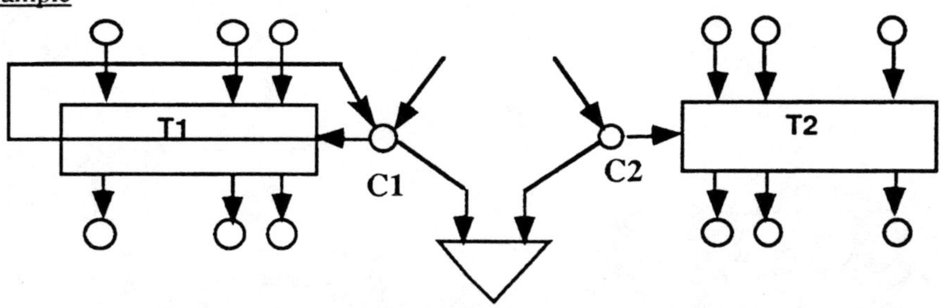

Let us create a SINK whose two inputs are also C1 and C2. Then, if both transitions are enabled (which means that tokens arrive simultaneously on C1 and C2, or a token arrives on C2 while another was looping on C1, waiting for inputs on transistion T1) the SINK is also ready to fire.

An error is detected on both C1 and C2 for being used in two transitions at the same time.

If we want to prevent only simultaneous firing of T1 and T2, the figure complicates a little bit :

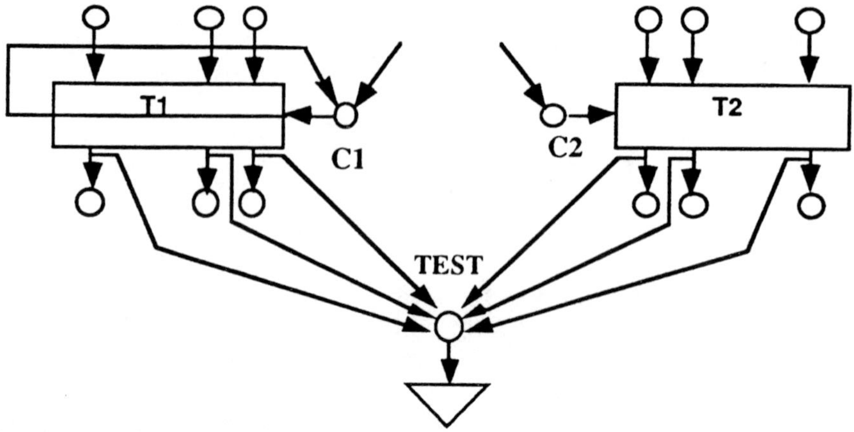

If T1 and T2 fire simultaneously, an error will be detected on place TEST for multiple loading.

c6 BRANCH

- INDEX is an integer

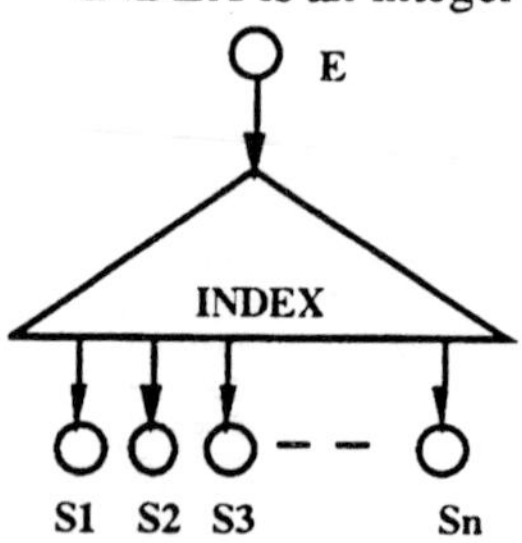

if INDEX = i, when a token arrives in E, then Si <- token, E <- empty

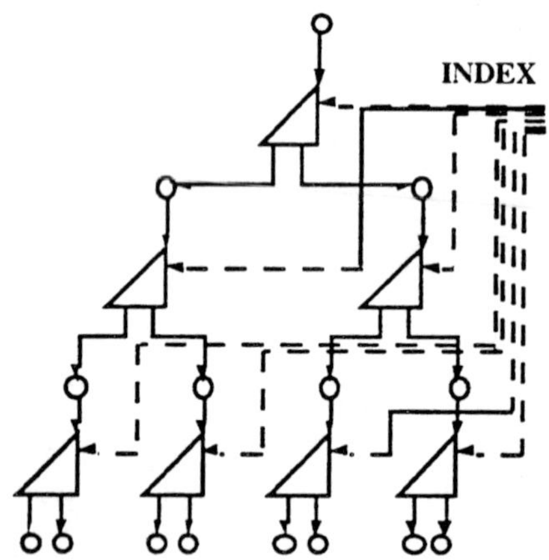

- INDEX is a vector of booleans :

A possible equivalent graph can be constructed with TEST primitives.

It is always possible through decoding to go back to this case

c7 Synchronized BRANCH

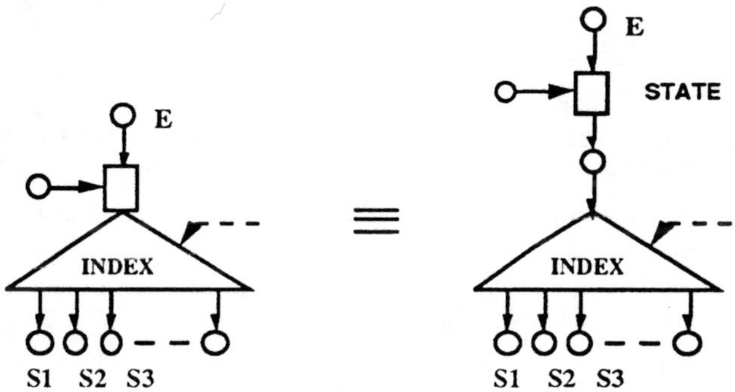

Example : FSM

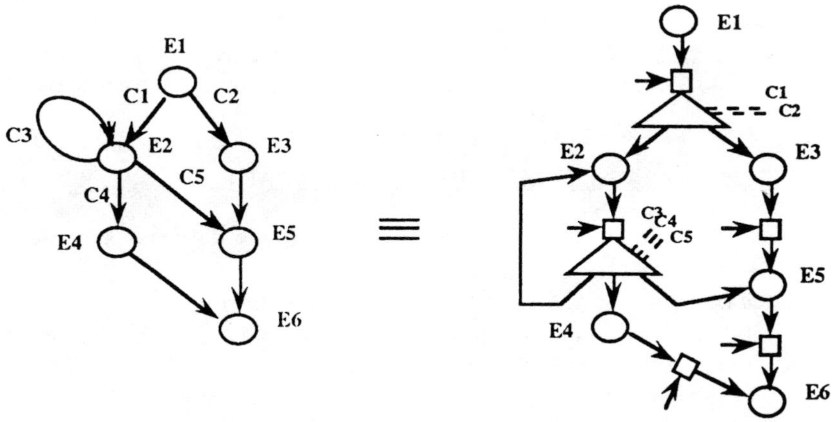

Usual state diagram

Equivalent control graph

If H is declared "clock", all occurences of 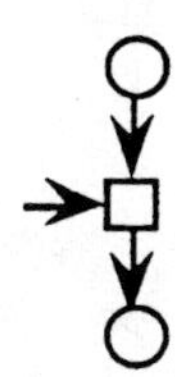 are equivalent to :

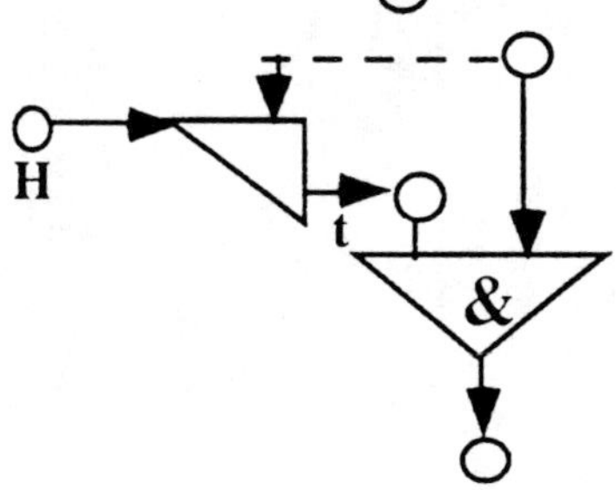

2. Principles of VHDL translation

TRANSITION FIRING

A transition firing means the occurence of one or more events (represented by tokens loaded in its input places) creating other events (tokens moved to output places).

When the "firing" occurs then input places that caused the transition firing are reset to *empty*.

Output places that stand for the newly generated events have their value set to *token*.

OPERATION STATEMENTS

An architecture of a described entity contains instructions and processes. They can be attached to a transition.

When this transition fires, the associated processes are activated through a "wait until <CONDITION>" instruction. CONDITION is a boolean set to true by the transition firing.

PRINCIPLES OF TRANSLATION OF THE CONTROL GRAPH

For each transition of the graph, a guarded block TRi and a boolean signal FIRE_TRi are created.

When the transition fires, the guard FIRE_TRi is set to "true"

The control is represented by the guarded block which describes the move of the tokens of the described transition when it fires.

The operative part which depends upon this control will be one or several processes sensitive to the boolean FIRE_TRi.

EXAMPLE OF AND TRANSITION WITH DELAY

Graphical representation

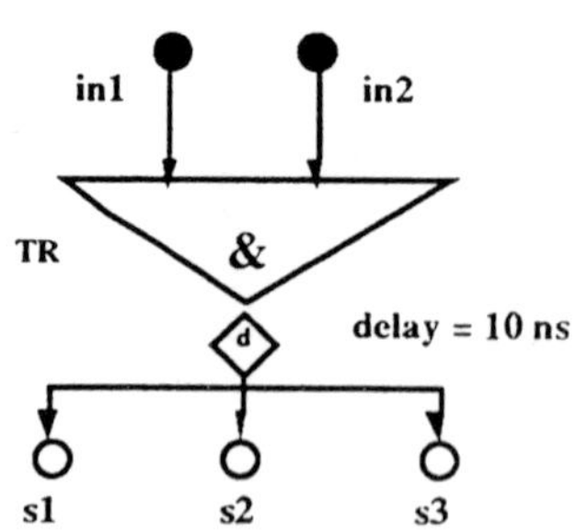

VHDL translation

```
FIRE_TR <= IN1=token and IN2=token ;
TR : block (FIRE_TR ) ;
        begin
                IN1 <= guarded empty ;
                IN2 <= guarded empty  ;
                S1  <= guarded token after 10ns ;
                S2  <= guarded token after 10ns ;
                S3  <= guarded token after 10ns ;
        end block TR ;
```

TEST TRANSITION

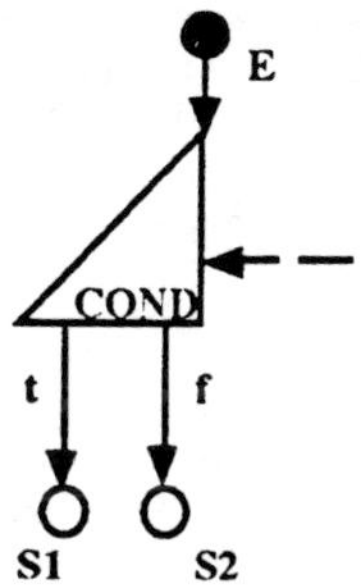


```
FIRE_TR <= E=token ;
TR : block (FIRE_TR ) ;
    begin
        E <= guarded empty ;
        S1  <= guarded token when
        COND=true else S1;
        S2  <= guarded token when
        COND=false else S2 ;
    end block TR ;
```

INDEX TRANSITION

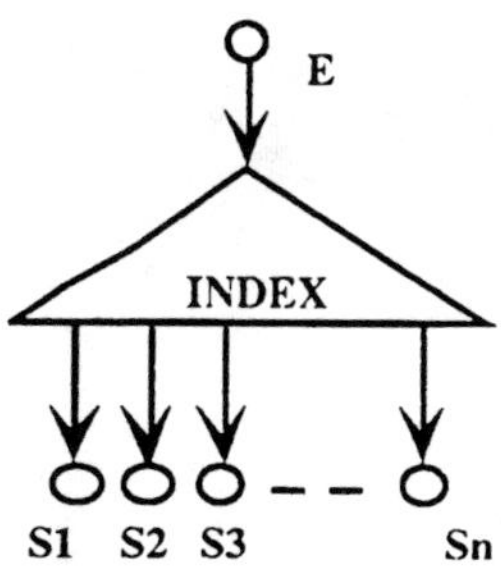


```
FIRE_TR <= E=token ;
TR : block (FIRE_TR ) ;
        begin
        E    <= guarded empty ;
        S1   <= guarded token when INDEX=1
                    else S1;
        S2   <= guarded token when INDEX=2
                    elseS2 ;
                    ...
        Sn   <= guarded token when INDEX=n
                    else Sn ;
        end block TR ;
```

SELECT TRANSITION

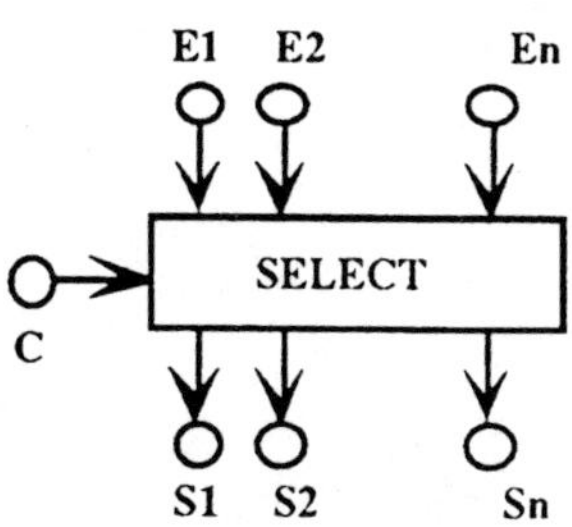


```
FIRE_TR <= (C=token and (E1=token
        or E2=token or ... or En=token) ) ;
TR : block (FIRE_TR ) ;
  begin
        TR1: block ( guard and E1=token);
                S1  <= guarded token ;
                C <= guarded empty;
        end block TR1;
        TR2: block ( guard and E2=token);
                S2  <= guarded token ;
                C <= guarded empty;
        end block TR2;
                ...
        TRn: block ( guard and En=token);
                Sn  <= guarded token ;
                C <= guarded empty;
        end block TRn;
  end block TR ;
```

PRIORITY TRANSITION

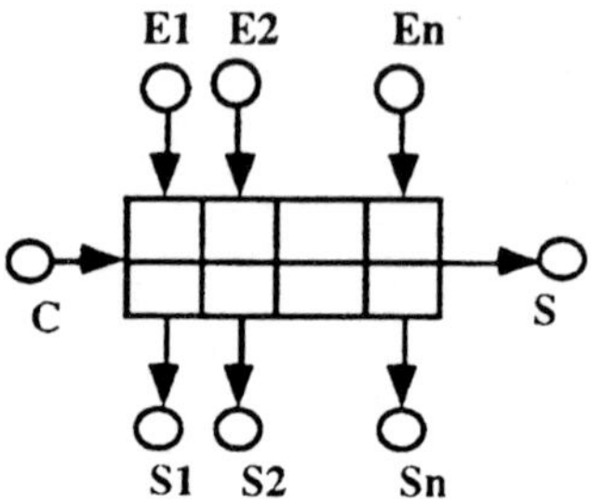


```
BTR1 <= TRUE ; BTR2 <= BTR1 and E1=empty;
                         BTR3 <= BTR2 and E2=empty;
                    ...
                    BTRn+1 <= BTRn and En=empty;
FIRE_TR <= (C=token ) ;
TR : block (FIRE_TR ) ;
  begin
      TR1: block ( guard and E1=token and BTR1);
              S1  <= guarded token ;
              C   <= guarded empty  ;
      end block TR1;
      TR2: block ( guard and E2=token and  BTR2);
              S2  <= guarded  token ;
              C   <= guarded empty  ;
      end block TR2;
              ...
      TRn: block ( guard and En=token and BTRn);
              Sn  <= guarded  token ;
              C   <= guarded empty  ;
      end block TRn;
      TRn+1: block ( guard and BTRn+1);
              S  <= guarded  token ;
              C  <= guarded empty  ;
      end block TRn+1;
  end block TR ;
```

Resolution function and package types

The type used for the places will be an enumerated type with values empty, token and undefined. As the places appear several times in the left part of loading instructions, a resolution function is necessary. This function will be used for detecting multiple loading. The type and resolution function will be defined in a special package:

```
package types is
   type States is (empty, token, undefined);
   type a_states is array (natural range <> ) of States;
   function token_ f (inp : a_states) return states ;
end types ;
package body types is
      function  token_f (inp : a_states)  return states is
      begin     assert (inp'length <= 1) report "loading conflict" severity warning;
                  if  inp'length <= 1 then          return inp(inp'low);
                  else  return empty;     end if;
      end token_f;
end types;
```

3. An example of VHDL translation

CONTROL GRAPH

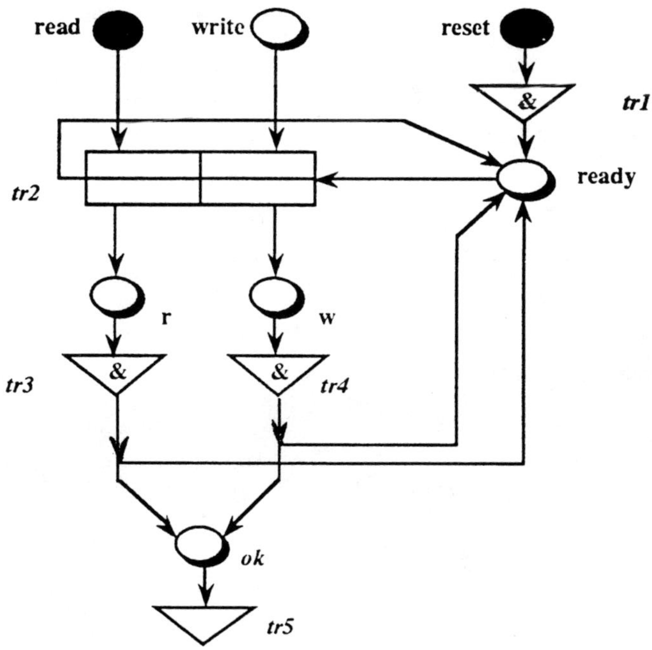

This graph represents the control part of a RAM memory. This part has three inputs and one output of type "token" and five outputs of type "boolean" which correspond to the firing of each of the five transitions.

The read and write operations cannot occur simultaneously. Then, when there are two tokens in the inputs, the read operation is executed first. The write token is memorized until read is reset to empty and ready to token.They can be performed only if the place "ready" has been reset.

When executed both actions issue an acknowledgement "ok" and reset "ready".

VHDL translation

entity RAM_LASSO is
end RAM_LASSO;

```
use work.types.all;   -- type and resolution function package
architecture CONTROL_GRAPH of RAM_LASSO is
signal read_ck, write_ck, reset_ck : bit;
signal BTR1,BTR2,BTR3 : boolean;
signal FIRE_TR2,FIRE_TR3,FIRE_TR1,FIRE_TR4,FIRE_TR5:
  boolean;
signal R,W,OK,READ,WRITE,READY,RESET :
          token_f  States register;
begin
  bl_read : block ( not read_ck'stable and read_ck = '1')
          begin   read <= guarded token ;      end block;
  bl_write : block ( not write_ck'stable and write_ck = '1')
          begin    write <= guarded token ;   end block;
  bl_reset : block ( not reset_ck'stable and reset_ck = '1')
          begin    reset <= guarded token ;   end block;

BTR1 <= TRUE ; BTR2 <= BTR1 and (READ=empty);
BTR3 <= BTR2 and (WRITE=empty);
FIRE_TR2 <= (READY=token) ;
TR2 : block (FIRE_TR2 )
   begin
      TR21: block ( guard and READ=token and BTR1)
       begin    R   <=  guarded token ;
                READY    <=  guarded empty ; READ <= guarded empty ;
       end block TR21;
      TR22: block ( guard and WRITE=token and  BTR2)
       begin    W   <=  guarded  token ;
                READY    <=  guarded empty ; WRITE <= guarded empty ;
       end block TR22;
   end block TR2 ;
FIRE_TR3 <= R=token  ;
TR3: block (FIRE_TR3)
      begin     R <= guarded empty ; OK   <= guarded token  ;
                READY   <= guarded token ;
      end block TR3 ;
```

```
FIRE_TR4<= W=token ;
TR4: block (FIRE_TR4 )
    begin   W <= guarded empty ; OK   <= guarded token ;
            READY   <= guarded token ;
    end block TR4;
FIRE_TR5<= OK=token ;
TR5: block (FIRE_TR5)
    begin   OK <= guarded empty ;
    end block TR5 ;
FIRE_TR1 <= RESET=token ;
TR1 : block (FIRE_TR1)
    begin   RESET <= guarded empty ; READY   <= guarded token ;
    end block TR1;
end CONTROL_GRAPH;
```

Bibliography

[LRM] Manuel de référence VHDL, IEEE std 1076-1987

[Ar89] J. R. Armstrong "chip-level modeling with VHDL", Prentice-Hall, 1989

[BL85] D. Borrione, C. Le Faou "Overview of the CASCADE multilevel hardware description language and its mixed-mode simulation mechanisms". Proc. of CHDL'85, Tokyo, Aug. 1985.

[Bo88] D. Borrione : "An approach to the formal verification of VHDL descriptions", Research Report n°683-I, IMAG/ARTEMIS, Nov. 1987

[Me85] J. Mermet "Several steps towards a circuit integrated CAD system: CASCADE", Proc. of CHDL'85, Tokyo, Aug. 1985.

[Mer85] J. Mermet "CERES Project - Final report" PRM-IMAG-PR-009-30, Oct. 1985.

[Bo87] D. Borrione : "LASSO user manual" Grenoble, IMAG/ARTEMIS, Jan 1987, RT 18.

[Ro72] C.W. Rose "LOGOS and the Software Engineer", Proc. Fall Joint Computer Conference, 1972, pp. 311-323.

[SB90] A. Salem, D. Borrione "Automatic formal verification of VHDL descriptions a first prototype", RR N° 823-I, IMAG/ARTEMIS, Grenoble, France, June 1990.

[Ra91] F. Rammig "Approaching System Level Design" Euro VHDL 91

[FIS64] Falkoff, Iverson, Sussenguth "Formal description of system 360" IBM Syst. Journal. Vol 3 pp 198-262, 1964